WEAPON

EARLY MILITARY RIFLES

BALÁZS NÉMETH

Series Editor Martin Pegler
Illustrated by Johnny Shumate & Alan Gilliland

OSPREY PUBLISHING
Bloomsbury Publishing Plc
Kemp House, Chawley Park, Cumnor Hill, Oxford OX2 9PH, UK
1385 Broadway, 5th Floor, New York, NY 10018, USA
E-mail: info@ospreypublishing.com
www.ospreypublishing.com

OSPREY is a trademark of Osprey Publishing Ltd

First published in Great Britain in 2020

A catalogue record for this book is available from the British Library.

ISBN: PB 9781472842312; eBook 9781472842329; ePDF 9781472842299; XML 9781472842305

20 21 22 23 24 10 9 8 7 6 5 4 3 2 1

Index by Rob Munro
Typeset by PDQ Digital Media Solutions, Bungay, UK
Printed and bound in India by Replika Press Private Ltd.
Osprey Publishing supports the Woodland Trust, the UK's leading woodland conservation charity.

Abbreviations

GOR	*General Orders and Regulations of the Army*
HKR	*Akten des Wiener Hofkriegsrates* (Files of the Imperial War Council in Vienna)
IKBPJCB	*Instruktion über die Kenntniß und Behandlung der Preußischen Jäger-Corps Büchse*
KA	Kriegsarchiv, Vienna
OM	*Ordnance Manual*
RER	*Regulations for the Exercise of Riflemen and Light Infantry*
RIS	*Regulations for the Inspection of Small Arms, 1823*
RKPI	*Reglement für die Königlich Preußische Infanterie*
RRC	*Regulations for the Rifle Corps formed at Blatchington Barracks under the Command of Colonel Manningham*

Acknowledgements

I would like to offer thanks for all the help I received in support of my research from Colonel Tamás Csikány of the Hungarian Academy of Sciences. Without his help my research would have been much more difficult. I would also like to thank the Early Firearms Department of the Museum of Military History, Budapest, especially Major Péter Soós, deputy director, and László Gondos, curator, for their assistance. Special thanks are due to my fellow YouTuber friend, Ian McCollum from Forgotten Weapons, and to Joel R. Kolander at Rock Island Auction for their kind help with the images. Special thanks are offered to Balázs Lázár of the Hungarian archival delegation at the Kriegsarchiv, Vienna, for his invaluable help in finding valuable sources. Special thanks are due to the teams at the Royal Armouries, Leeds, UK; Springfield Armory, Massachusetts, USA; Dorotheum GmbH, Vienna, Austria; College Hill Arsenal, Nashville, Tennessee, USA; and Morphy Auctions, Denver, Pennsylvania, USA. Great thanks to my friend, shooting and collecting buddy János Balassi, for lending me his *Jäger* rifles for photography. Last but by no means least, my greatest thanks go to my family – my wife Brigitte, my daughter Fanni, my son Sebastian, my parents and my brother – for their support, and for humouring me each time I start talking about old guns.

Editor's note

The *Gulden* or *forint* (fl.) was the currency of the Habsburg Empire; it was divided into 60 *Kreuzer* or *krajczár* (kr.). With the exception of calibres and powder weights (given in grains and in grams), metric units of measurement have been used in this book.

Common units of measure 1816–40

	Prussia	Austria	USA/Britain	France
Length				
Punkt / Point	0.182mm	0.1829mm	not used	not used
Linie / Line / Ligne	2.18mm	2.195mm	2.54 or 2.11mm	0.232cm
Zoll / Inch / Pouce	2.616cm	2.634cm	2.54cm	2.78cm
Fuß / Foot / Pied	31.39cm	31.61cm	30.48cm	33.33cm
Schritt / Pace	75cm	75cm	76cm	not used
Weight				
Gran / Grain / Grain	0.061g	0.07g	0.0648g	0.0541g
Loth / Loth	14.616g	17.5g	not used	not used
Drachm / Dram / Gros	3.65g	4.37g	1.772g	3.9g
Uncia / Ounce / Once	29.23g	35g	28.35g	31.25g
Pfund / Pound / Livre usuelle	467.711g	560.012g	453.6g	500g

Front cover, above: This Ferguson rifle was made by Durs Egg in about 1776. (© Royal Armouries XII.11209)
Front cover, below: © Osprey Publishing.
Title-page illustration: This illustration by Rudolf Ottenfeld, published in 1895, shows a Habsburg Army *Stutzen-Jäger* in 1809. (Author's collection)

CONTENTS

INTRODUCTION

This Pennsylvania flintlock long rifle with raised-relief carved stock and brass lock plate was made by Adam Ernst of York County, Pennsylvania. The American rifleman and his long rifle became symbols of freedom, even if their role in the American Revolutionary War (1775–83) was subsequently exaggerated. The American flintlock rifle of the 18th century was not able to replace the smoothbore musket as a bayonet could not be mounted on the rifle's barrel and the slim stock was too fragile for close combat. The early history of the long rifle is hard to research as the early gunmakers did not mark their products. It is not an easy task to identify these rifles as they could be very similar to European rifles. The use of commercial gun parts imported from Europe also makes identification more difficult. One of the features that aids identification is the hinged patch box – European *Jäger* rifles were usually equipped with wooden sliding patch boxes. The transition to a hinged, brass lid patch box was clearly a development of the colonial gunsmiths (Kauffman 2005: 8–15). (Rock Island Auction)

The rifle was not born as a military arm. In fact, until the mid-19th century, only a small proportion of infantry troops were armed with rifled firearms capable of long-range accuracy, even though rifled firearms had been widely used by civilians for hunting and target shooting all over Europe since the 17th century. The circumstances surrounding the invention of rifling remain obscure (the stabilizing effect of a spinning projectile fired from a bow or crossbow was well known before the invention of firearms), but this development is possibly connected to South German territories and can be traced to the end of the 15th century. The first rifling had straight grooves cut into the bore to give more space to the extensive residues of black powder. Later, these grooves were cut in spiral forms to spin the tight-fitting ball.

The history of the military rifle is closely connected to the history of the development of regularized light infantry. The first documented rifle-armed troops were the bodyguards of King Christian IV of Denmark (r. 1588–1648), serving during 1611–22. The wider picture of the evolution of light troops is quite diverse. Rifle-armed auxiliaries such as irregulars, militia and sharpshooters played only a secondary role, involving small raids, reconnaissance, picket duty and other traditional light-infantry tasks. Regularizing such forces was an important issue in the late 18th century. The archetype of the light rifle troops originated in Prussia. King Frederick II (r. 1740–86) raised his 700-strong rifle corps – the *Feldjäger-Korps zu Fuß* – in 1741 during the War of the Austrian Succession (1740–48). Although this unit was primarily intended to be an internal security force, it paved the way for regularized light infantry, which became an essential part of the Prussian Army; others followed suit, and in many armies the rifle-armed troops soon became the élite core of the light infantry.

This work covers the development of rifles and regularized rifle troops used by four, later five, armies – those of Prussia, the Habsburg Empire,

FAR LEFT
A Prussian *Jäger* in 1773 is depicted in this illustration from Gumtau 1834. (Author's collection)

LEFT
A Habsburg Army *Jäger* during the Seven Years' War (1756–63) is depicted in this coloured lithograph by the Austrian artist Franz Gerasch (1826–1906). (Author's collection)

the United States, Britain and latterly France – that all chose different paths to achieve the same solutions, while contributing much to the development of rifle tactics, and provided examples for others to follow. The Kingdom of Prussia and the Habsburg Empire created the archetype of the regularized rifleman, who was selected from the best soldiers of the line-infantry regiments and received the line-infantry drill before being further trained to become a rifleman, joining the élite of the army's regular troops. The United States' military rifle culture had a profound impact on the organization and fighting methods of the citizen armies of the French Revolutionary Wars (1792–1802) and the Napoleonic Wars (1803–15), and an unquestionable influence on the modernization of infantry tactics. Conversely, Britain outsourced rifle culture for many years by recruiting light troops from German states, only later organizing its own rifle troops in earnest. France largely eschewed rifles until the 1820s, when French inventors played a major part in the development of new rifle technologies and tactics.

This work provides a detailed look at the development of the military rifle firing a patched round ball and its practical use; the organization, training and tactics of the rifle-armed infantry troops; and the challenges of rifle development and production, including the struggles to implement interchangeability in firearms manufacturing. It also offers a unique insight into the difference between the smoothbore musket and the rifle in terms of ballistics and handling with the help of an experimental archaeology project involving original firearms of the early 19th century.

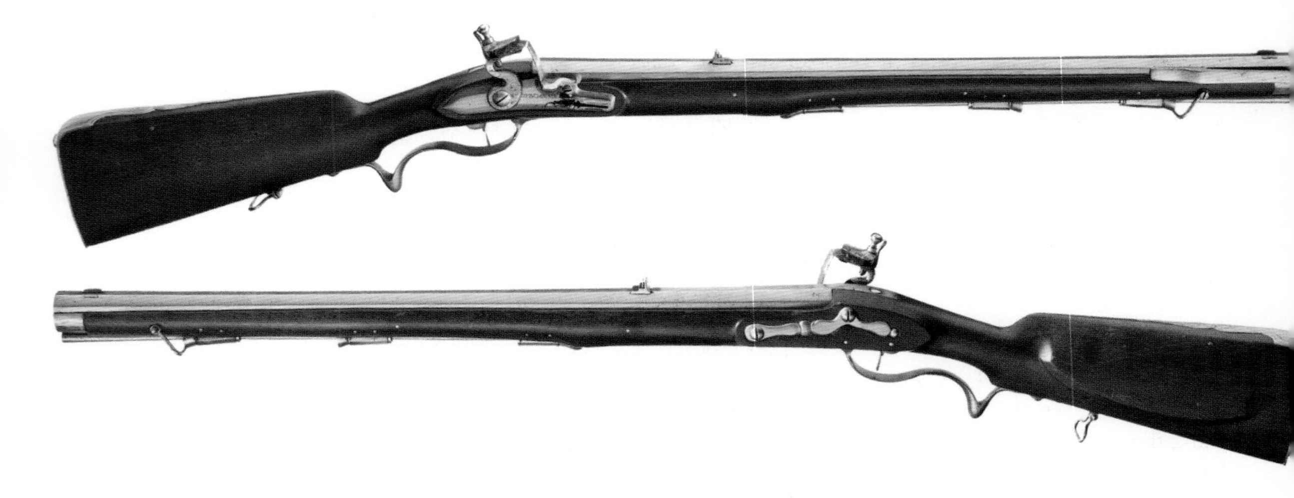

DEVELOPMENT

The evolving rifle

A Prussian *Alte Corpsbüchse* from the second half of the 18th century. As no central patterns existed for manufacturing, many variations exist. These rifles were not originally equipped with bayonets. The first recorded reference to attaching a bayonet to the Prussian military rifle was in November 1784, when Hauptmann von Böltzig suggested their use (Gumtau 1831: I.76). (Image courtesy of Hermann Historica)

The key elements of the early military rifle, the flintlock mechanism and the rifled bore, existed in the early 18th century: the question was how to use such weapons effectively in war. At the outset of our period, muzzle-loading smoothbore flintlock firearms ruled the battlefields of Europe and North America. The first 90 years of the military rifle would be dominated by this technology, with only a handful of innovative breech-loaders pointing the way ahead. Percussion ignition would play a major role in military rifle development from the 1830s, as the specialist military rifle finally evolved into a weapon that could be issued to all soldiers.

RIFLES OF THE FLINTLOCK ERA

In the 18th century, standing armies deployed rigidly drilled infantry armed with smoothbore muskets fitted with socket bayonets, and utilized close-order linear tactics. The smoothbore muskets were inaccurate and the drill lacked any marksmanship training, so firing volleys at the large target area of the enemy's close-order combat formations was the only effective way to fight. At first, rifles played only a marginal role on campaign, in the hands of auxiliaries who conducted the 'small war' (also known as the partisan war, *kleiner Krieg*, *guerrilla* or *petit-guerre*); they took little part in the pitched battles that decided conflicts such as the Seven Years' War (1756–63). The rifles carried by such troops, many being civilian weapons belonging to the soldier himself, reflected their specialist roles.

The American and French Revolutionary wars transformed the nature of land warfare. In the struggle for North America, accurate fire from rifle-armed Patriot irregulars took a heavy toll on British close-order infantry formations, prompting the British to adapt their own tactics and revive their light-infantry arm. A decade later, faced with the fast-moving attack columns and open-order skirmish lines employed by the poorly trained but highly motivated citizen soldiers of Republican France, the armies of Prussia, the Habsburg Empire, Britain and others strove to develop their own light infantry in response. These developments increased the need for accuracy and promoted the value of the military rifle. Flintlock rifle technology and tactics would reach their perfected form during the Napoleonic Wars.

Prussian flintlock rifles

The *Jäger* raised in 1741 were selected from hunters, already accustomed to using the rifle. Their firearm was the *Jägerbüchse* or *Corpsbüchse*, a short, large-calibre hunting rifle firing a tight-fitting, patched round ball. Little information survives about these rifles, as up until the acceptance of the M1810 (or M1811 as it is also named) *Neue Corpsbüchse* there were no central patterns in Prussia for manufacturing military rifles.

Soon after the accession of King Frederick William II (r. 1786–97) the light capabilities of the line-infantry regiments were increased by selecting ten sharpshooters, or *Schützen*, from each line-infantry company to fight in open order. These soldiers were intended to provide a 'screen' in front of the deployed line when advancing or retreating. Their armament was the M1787 *Schützengewehr*. This rifle was much shorter than the infantry musket (124cm rather than 145cm) and in a smaller calibre (18.5mm bore diameter rather than 19.6–20.4mm).

In the wake of Prussia's disastrous war against Napoleonic France in 1806–07, the Treaties of Tilsit (7 and 9 July 1807) had a profound impact on the country. Determined to launch a full-scale reform of the Army, King Frederick William III (r. 1797–1840) ordered the activation of the Militär-Reorganisationskommission (Military Reorganization Commission) in July 1807. Chaired by Generalmajor Gerhard von Scharnhorst, the Commission was tasked first with evaluating the causes of the Prussian defeats, and second, with proposing reforms to transform the Prussian Army by developing a system of recruitment, military education and training. Scharnhorst's Commission immediately started to work on reform of the infantry armament. Based on the French M1777 musket, a new general-service infantry arm, the 18.5mm-calibre 'Neupreußisches' or 'Scharnhorst'sches' M1809 infantry musket, was accepted on 19 May 1809. The new rifle for the light troops arrived only a year later when, following the Commission's testing of the M1810 *Neue Corpsbüchse* in the autumn of 1810, that weapon was accepted for service; the first M1810 rifles arrived from the Potsdam factory on 1 July 1811 (note that the M1810 is sometimes designated the M1811).

The M1787 *Schützengewehr* was issued with a socket bayonet and a pole to support the rifle for accurate shooting. Featuring a block rear sight for shots up to 150 paces (113m) and a folding rear sight for shots up to 300 paces (225m), the *Schützengewehr* was intended to be fired as quickly as possible. Leather padding was supplied to all soldiers to save their left hands from the overheating barrel. (© Royal Armouries XII.2281)

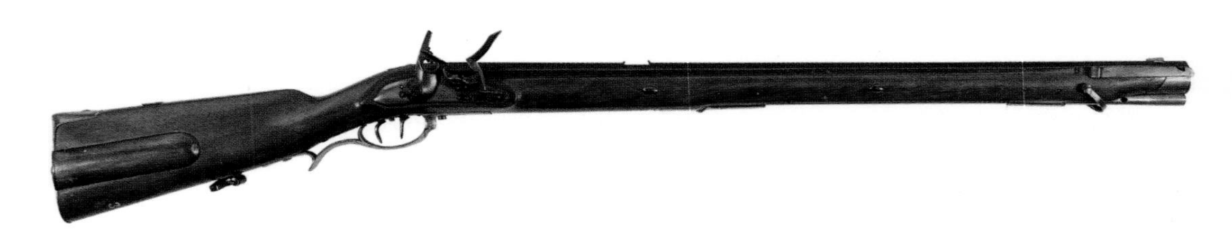

This Potsdam-manufactured M1810 *Neue Corpsbüchse* lacks the jaw screw and rear sight. All M1810 rifles had a French-pattern lock, with a brass pan and reinforced hammer with a heart-shaped opening. The rifle featured a typical German full stock with a patch box in the butt stock, an iron ramrod placed in the channel of the forestock, a bayonet rail attached to the right side of the barrel, a bead front sight and a folding rear sight with two leaves. The rifle was equipped with a double-set trigger system: the rear trigger served to set the mechanism and the front trigger released the shot with minimal trigger-pull weight. (© Royal Armouries XII.2282)

Although one of the Commission's intentions was to standardize existing Prussian Army rifles, in practice the new firearms served only as a guideline. Up to the end of the M1810's production period in 1835 there were significant differences between the rifles manufactured in individual facilities. The first samples were made in Potsdam, but later three other factories in Neiße, Saarn and Suhl were also involved in the production. The type of M1810 rifle produced by the different factories varied in many specifications (Gumtau 1835: II.23–29).

Specifications of the M1810 rifles made in different locations

Factory	Number produced to 1831	Total length	Calibre	Weight	Grooves	Barrel length	Rifling twist-rate	Rifling profile	
								Width	Depth
Potsdam	3,848	113cm	14.64mm	4.6kg	8	72.7cm	1:48.5cm	2.6mm	0.5–0.7mm
Neiße	668	Same as the Potsdam-made rifle							
Saarn	416	109cm	14.64mm	4kg	7	62cm	1:46.65cm	3.4mm	0.5–0.7mm
Suhl	1,113	107cm	14.64mm	3.88kg	7	68cm	1:51cm	n/a	0.5-0.7mm

Habsburg flintlock rifles

Up until 1815 the Habsburg Army utilized both regular and irregular (*Frei-Korps*) troops for light-infantry duty, but both troop types were always disbanded or merged into regular units in peacetime (Nagy 2013: II.131). The first regular *Jäger* troops were raised in 1759, following the Prussian example in the Seven Years' War. Many of the soldiers, recruited from hunters, carried civilian hunting rifles. In 1769 a larger *Jäger* force of ten companies was created. From the mid-18th century, the eastern and southern borders of the Habsburg Empire were protected by *Grenz-Infanterie-Regimenter* (border-guard infantry regiments). Each company had 20 sharpshooters named *Grenz Scharfschützen*.

The first rifled arm of the light troops of the Habsburg Army was the M1754 *Jägerstutzen*, the archetype of the short German military rifle. The calibre was 1 Loth (15mm), with a 79cm barrel. A wooden ramrod fitted into a channel on the underside of the stock, while the bore was rifled

The M1768 *Doppelstutzen* rifle had such a large touch hole that the powder from the main charge poured into the closed pan when the rifle was being loaded. Loading was faster, but much gas escaped through the large hole, resulting in inaccuracy when fired. Although the *Doppelstutzen* rifle was considered for general infantry issue, its substantial weight (5.35kg) and high cost of manufacture prevented this. (Author's collection)

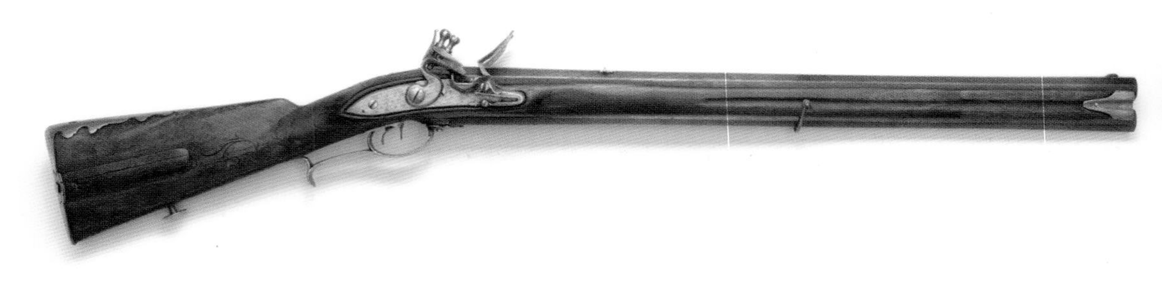

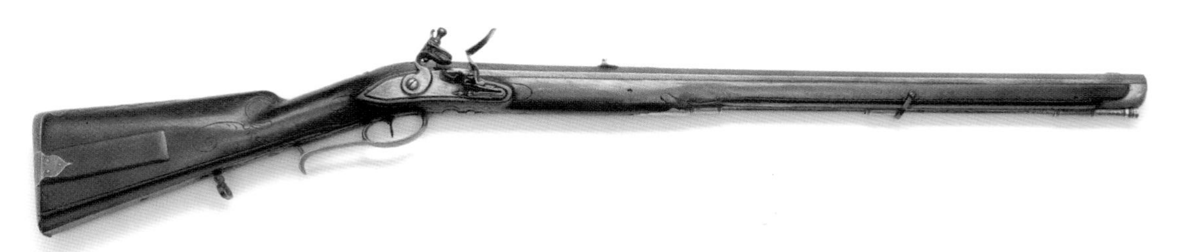

The M1769 *Jägerstutzen* rifle had a 14.6mm-calibre barrel, rifled with seven grooves and lands, a block rear sight (*Standvisir*) and a brass dovetailed front sight. The twist-rate of the rifling was one turn in 890mm – quite fast compared to that of American or British rifles but slower than that of Prussian rifles of the period – and the length of the barrel was 670mm (Götz 1978: 177). The ramrod was made of iron. No bayonet could be attached to the rifle; instead the *Jäger* carried a *Hirschfänger* ('deer catcher'), a straight double-edged hunting knife, for close combat (Gabriel 1990: 38). (Author's collection)

with seven grooves and lands. The total length of the rifle was 112cm, with a weight of 3kg, and it was issued to the Habsburg Army's newly raised *Jäger* units and to the *Grenz Scharfschützen* (Dolleczek 1970: 77).

The M1768 *Doppelstutzen* rifle was designed for the selected *Grenz Scharfschützen* (Dirrheimer 1975: 56) and was one of the first rifles (the other being the M1769 *Jägerstutzen*) to be manufactured according to central patterns. Each M1768 had two 65cm barrels in 14.8mm calibre, one rifled with seven grooves for long-range accuracy, and the other smoothbore for rapid fire. The design went through minor modification during the firearms reforms of the 1790s, resulting in the M1795. The iron pan of the lock was replaced with a detachable, tilted brass pan for easier loading, and the hammer was strengthened. The *Doppelstutzen* rifles were withdrawn from service in 1805–07 but their locks were later used on *Jägerstutzen* rifles.

The number of rifle-armed regular light infantry in Habsburg Army service had increased by the end of the 18th century, but the first years of the French Revolutionary Wars revealed the obsolescence of their firearms and military structure. The Habsburg monarch – who was Holy Roman Emperor (1792–1806) as Francis II, and Emperor of Austria (1804–35) as Francis I – and his brother, the Archduke Charles, agreed that reforms were necessary, but the Archduke did not agree that firearms reform was practical while the Habsburg Empire was in such a poor state. The Emperor was insistent, however, and ordered the creation of a *Militär-Hof-Commission* (independent military commission) to conduct the reforms. Led by Baron Joseph Alvinczy von Borberek and excluding the Archduke Charles, the Commission held its first meeting in 1798. Leopold von Unterberger, a renowned military engineer, was in charge of firearms. The key elements of Unterberger's reforms were: the adoption of the principle of interchangeability; the reduction of the musket calibre to the 17.5mm French calibre; the reduction of the musket ball to a diameter of 15.9mm; and the change from a self-priming system to a conical touch hole in order to achieve better accuracy, thereby sacrificing the high rate of fire of the old muskets (Unterberger 1807: 14–25).

The new series of firearms to be adopted included the M1798 infantry musket for the Habsburg Army's line, light and *Grenzer* infantry, the M1798 long carbine for the dragoon regiments, the M1798

Austrian M1769 *Jägerstutzen* rifle modernized in 1795–96. The original M1769 rifles had full stocks reaching the muzzle; there was no place to attach the bayonet. In 1795–96 many of them were modified to accept the new 66cm sword-bayonet. This model was also equipped with a double-set trigger mechanism. (Author's collection)

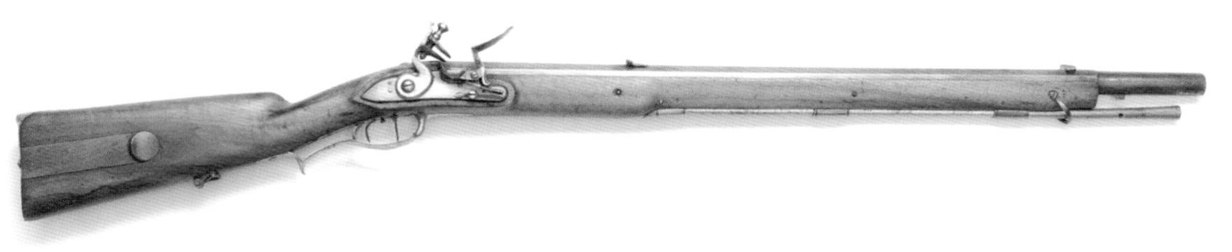

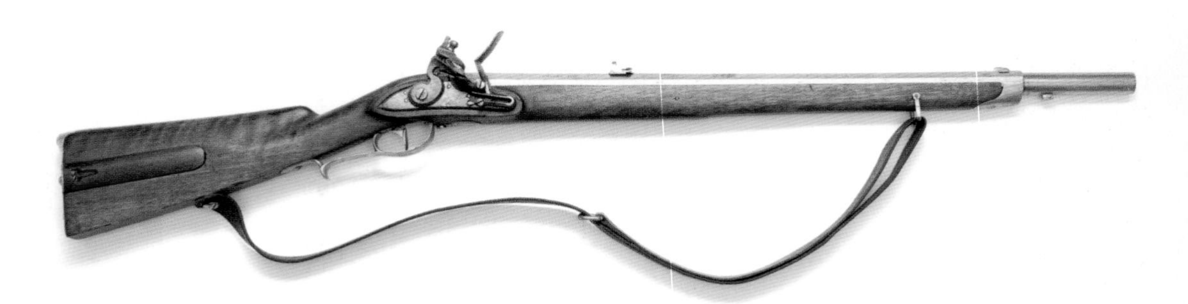

Its lock clearly showing the influence of the French M1777 smoothbore musket, the M1807 *Jägerstutzen* rifle was a minor modification of the M1795/96 rifle. It featured a tilted brass French-style pan, the reduced-bore calibre of 13.9mm and the new leaf rear sight, and it accepted a socket sword-bayonet. (Author's collection)

short carbine for the hussars and other light-cavalry units, the M1798 cavalry rifle for the mounted *Jäger*, the M1798 cavalry pistol, and the M1795 *Doppelstutzen* and M1795/96 *Jägerstutzen* rifles. Production of the new-pattern weapons started in state-owned factories in Vienna, Steyr and Hainfeld, but after a few years many private companies in Ferlach, Prague, Carlsbad, Wisenthal, Warnsdorf, Weippert, Pottendorf, Wiener Neustadt, Neuberg, Brescia and Lomezano were also contracted. Owing to the difficult economic situation the Habsburg Empire faced, the complete rearmament of the Army took nearly ten years (Dolleczek 1970: 80–81).

The M1795/96 *Jägerstutzen* rifles were the first modernized Austrian firearms based on combat experience fighting the French and were an updated version of the M1769 with a few modifications. The straight steel pan was replaced with a tilted brass pan on the lock and the last 11cm section of the barrel was reshaped round to accept the new socket-type 66cm sword-bayonet (*Haubajonett*) which increased the close combat capabilities of the *Stutzen-Jäger*. The land-to-land diameter of the bore was 14.5mm, increasing in diameter towards the breech for better resistance to fouling, and the seven-groove rifling made a complete turn in 880mm. The total length of the rifle was 105cm, with a weight of 4kg (Götz 1978: 176–77). The service charge was 1 Quintel (4.375g, 67.5 grains) of *Scheibenpulver* (fine rifle powder), and a 14.16mm patched lead round ball was used. The butt stock featured a wooden sliding patch box for the maintenance tools, and a cheek piece on the left side for comfortable aiming (Unterberger 1807: 42).

The *Militär-Hof-Commission* was not satisfied with the calibre of the *Jäger* rifles and recommended following the Prussian example by increasing the ball diameter to 15.8mm. The idea was rejected, however, as the subsequently greater recoil worsened accuracy. The calibre was even reduced to 13.9mm on the Commission's recommendation, but the size of the projectile remained the same, resulting in a change in loading: the ball was larger than the bore, so it had to be hammered into the muzzle. The tight fit contributed to accuracy but reduced the rate of fire (Götz 1978: 178). New sights were added: a *Standvisir* (block sight) and a *Klappvisir* (leaf sight). The barrel was blued to counter corrosion and to avoid reflecting the sunlight. The trigger system was improved by adding a fly to the tumbler to reduce the trigger pull. (A fly is a small metal sheet connected to the tumbler with an axis; it aids the sear jumping over the half-cock notch on the tumbler, so the weight of the

EARLY HABSBURG BREECH-LOADING RIFLES

The first breech-loader to be adopted by the Habsburg Army was the M1770 carbine, designed by Giuseppe Crespi as a conversion of existing military muzzle-loaders. Emperor Joseph II (r. 1765–90) approved the design and ordered a few thousand smoothbore breech-loading carbines for various cavalry units. The M1770 had a hinged, tip-up breech section, but the closure was not gas-tight, so after a short trial period the weapon was withdrawn from service.

Although the first experiences were unsuccessful, the breech-loading concept was not forgotten. During the 1770s a firearms maker from Ampezzo, Bartholomäus Girardoni, continued the development of the Crespi concept. Girardoni's first breech-loading system worked on the same principle, but he automated the cocking of the hammer upon opening the breech. The operation needed only four motions: first, opening the breech; second, inserting the cartridge; third, closing the breech; fourth, firing the gun (Dolleczek 1970: 70). Girardoni also experimented with repeating systems by attaching a powder magazine to the right side of the barrel of a flintlock rifle and a bullet magazine to the left side. The heart of the repeating system was a vertically moving breech block that fed the powder and the bullet into the breech. Girardoni was not able to seal the breech properly, however, and his left arm was seriously injured when the powder magazine exploded. This failure led him to replace the black powder with a safer propellant: compressed air.

The air rifle (*Windbüchse*) was not a new invention, being widely used all across Europe for hunting from the 17th century onwards. Girardoni's repeating system closely resembled that of his repeating flintlock rifle, with a magazine for the bullets and a sliding breech block. The compressed-air bottle was screwed to the frame, and served as the butt stock of the weapon. The calibre of the rifle was 13mm, and its effective range was 150 paces (113m). The compressed-air bottle was replaceable and each soldier carried 2–4 of them in a specialized leather holder. The bottles were charged with a dedicated pump and the pressure was enough for 40 shots, but muzzle velocity lessened after the first 20 or so shots. Joseph II decided to adopt the M1780 *Repetierwindbüchse* for the Habsburg Army's sharpshooter units. The rifle remained in service until the end of the Napoleonic Wars, but it was deemed to be too complicated for the troops, so it was withdrawn from service in 1815.

In 1780, Bartholomäus Girardoni's M1780 *Repetierwindbüchse* was the first repeating air rifle ever accepted for military service. Four soldiers of each Habsburg Army *Jäger* company received an M1780 *Repetierwindbüchse*, and in 1790 it was decided to equip an entire new Tyrolean rifle unit – 1,313 men – with the weapon (Lugosi 1977: 11). Although it proved to be a formidable arm, it was removed from service in 1815 due to its complexity. (Museum of Military History, Budapest)

OPPOSITE TOP
A 1792 Contract flintlock rifle in .50 calibre. The cost of one rifle of the first batch of 1,000 was $12, which was quite low compared to the average cost of a good-quality long rifle. The metal parts – especially the locks – were in most cases purchased from importers. All military rifles were marked with the letters 'US' stamped on the barrel (Flanagan 2008: 33). (Photo courtesy of Morphy Auctions, www.morphyauctions.com)

trigger pull can be lightened to a few grams.) Each soldier received a brass powder measure (see page 47) for charging the bore with loose powder or for making paper cartridges (Paumgartten 1802: 72).

US flintlock rifles

The flintlock long rifle saw extensive combat in the hands of militiamen on both sides during the American Revolutionary War. The original German *Jäger* rifle from which such weapons originated had a relatively short (762–1,016mm) 'swamped' octagonal barrel (meaning that the barrel had a larger external diameter at the breech and the muzzle; removing material from the central section of the bore moved the centre of gravity closer to the shooter's supporting hand, thereby aiding a more comfortable hold of the rifle). Calibres were smaller than the musket ball, the most common being 13–16mm. The stocks were made from walnut or fruit wood and their length reached the muzzle. Such rifles often featured a double-set trigger system in which the rear trigger set the springs of the firing mechanism, while the front trigger fired the rifle with minimal force.

The development of the long rifle can be divided into three periods. The transition from the *Jäger* rifle to the Pennsylvania long rifle took approximately 50 years, roughly 1725–75. The outbreak of the American Revolutionary War in 1775 marked the beginning of the second period, which lasted until the beginning of the percussion ignition era (see page 23). The general shape and features of the Pennsylvania long rifle were perfected in the 25 years after 1775 (Kauffman 2005: 17). The major difference compared to *Jäger* rifles was the longer, small-calibre bore. The barrel remained octagonal and swamped but the calibre was reduced to 9–15.2mm. The barrel length was increased to 762–1,270mm, though the longer but smaller-diameter barrel did not need more iron than the robust *Jäger* barrels. The smaller calibre needed less lead and less powder – a saving of valuable raw materials in the New World – and the projectile propelled in the longer barrel also utilized better the energy of the charge, so the powder load could be further reduced.

Signed by the Pennsylvania gunsmith John Moll and dating from the first two decades of the 19th century, this flintlock long rifle features an intricately engraved patch box. The Pennsylvania-Kentucky rifle, or the American long rifle, owes its origins to the German *Jäger* rifle. The name is misleading, however, as these rifles were not only manufactured in Pennsylvania and Kentucky, but also in Maryland, New England, New York, North Carolina, Ohio and. Virginia. The name 'Kentucky rifle' was popularized by later collectors as the general designation for all American long rifles (Kauffman 2005: Preface). This rifle was the most important arm of the early American riflemen. (Rock Island Auction)

Passed after the US defeat at the hands of the Native Americans at St. Clair on 4 November 1791, the Militia Acts of 1792 required 'each and every free able-bodied white male citizen of the respective states, resident therein, who is or shall be of the age of eighteen years, and under the age of forty-five years' to enrol in the local militia company. Each man brought his own firearm, with all necessary accessories. To arm additional rifle troops, the US Army centrally purchased rifles that followed the general long-rifle design: 3,000 pieces were ordered to equip three newly raised rifle regiments. On 4 January 1792, Secretary of War Henry Knox ordered Major-General Edward Hand to contract Pennsylvania and Lancaster gunsmiths to produce flintlock long rifles with 42in (106.7cm) barrels in .50 calibre, and with a brass patch box and a fly in the tumbler. Between 1792 and 1794, 3,476 pieces were delivered to the US Government from Pennsylvania contractors such as Jacob Dickert, Peter Gonter and Jno. Graeff (Flanagan 2008: 30). All these rifles shared the same barrel length and calibre but there were many minor differences between the products of different makers.

Secretary of War Henry Dearborn expressed his dissatisfaction with the contract rifles of the US Army regular troops and on 25 May 1803 ordered Joseph Perkin, superintendent of the Harper's Ferry Armory, to design a new, improved military rifle. Dearborn demanded a shorter, larger-calibre, half-stocked rifle. It took Perkin just six months to submit several patterns, out of which the winning design was selected by the War

This image shows the Harper's Ferry Armory in 1861, at the outbreak of the American Civil War (1861–65). The rise of large-scale firearms manufacturing and the birth of the US-made standardized military rifle are closely related to the establishment of the national arsenals. Before 1794 all US military small firearms were purchased from contractors and importers, thus entailing a high dependency on commerce. To ease this problem the US Congress decided to establish four national armouries, and approved a bill for repairing existing arsenals and magazines. In the event there were two armouries: one at Springfield, Massachusetts, and the other a new foundation at Harper's Ferry, Virginia. Situated at the confluence of the Potomac and Shenandoah rivers, Harper's Ferry was a favourable location as iron, coal, wood and water power were nearby, and it was not far from central US Army forges and furnaces. The Harper's Ferry venture was supervised by Joseph Perkin, an Englishman working in the gun trade who had emigrated to the United States in 1774. Perkin's considerable knowledge and experience were suitable for the management of the new plant, so Secretary of War James McHenry ordered him to take his position as superintendent in 1798. The production of firearms at Harper's Ferry was initiated sometime in early 1800 (Smith 1977: 27–52). Springfield Armory was mainly responsible for producing muskets, while the new plant at Harper's Ferry focused upon rifle and pistol production in its early years. (Author's collection)

Department. In a letter to Perkin, Dearborn expressed his approval, requiring only minor changes to the design, including adjusting the shape of the upper thimble for the ramrod, widening the aperture in the sight near the breech and adding a brass ferrule near the tail pipe to strengthen the stock (Carrick 2008: 2). An order for 2,000 pieces was immediately placed, and was increased to 4,000 in November 1804. The plan was to deliver 2,000 pieces each year, but mechanical difficulties and an outbreak of malaria slowed production, delaying the final shipments until 1807 (Smith 1977: 54).

Tench Coxe, Purveyor of Public Supplies during the Jefferson administration (1801–09) and the person responsible for procuring all firearms for US service, was not satisfied with the M1803 at all. He requested a longer (38in, 96.5cm) barrel, a full stock, a brass tip at the end of the iron ramrod to save the rifling, and improvements to the

ABOVE & RIGHT
This first-production Harper's Ferry Armory M1803 flintlock rifle is dated 1805. Representing an excellent balance between the long rifles and the European *Jäger* rifle, the M1803 is one of the most elegant military rifles ever designed. The barrels of the first production batch were 33in long, octagonal at the breech and round towards the muzzle. The half-stock was made from walnut with a brass patch box. The calibre of the bore was .54in; the bore was rifled with seven spiral grooves with a twist-rate of one complete turn in 49in (124.5cm). The rifle featured an iron ramrod and a distinctive iron rib under the barrel, but the rifle was not equipped with a bayonet. (Rock Island Auction)

construction of the lock and trigger mechanism (Russell 2005). As production at Harper's Ferry had a slow start, the involvement of private contractors was necessary. The US Government sought contractors, paying $10 for each rifle – later named the 1807 Contract rifle – made with a .54-calibre barrel 38in (96.5cm) long. The profile of the barrel was to be one-third octagonal and two-thirds round towards the muzzle in order to save weight and achieve better balance. There were no other specifications given to the gunsmiths, and surviving specimens all follow the basic design of the Kentucky-Pennsylvania rifles. A total of 1,806 rifles were delivered by rifle makers in Lancaster, Pennsylvania, up until 1809 (Flanagan 2008: 34). The Militia Act of 1808 secured the necessary funds of $200,000 for purchasing and manufacturing military equipment for the state militias.

The War of 1812 (1812–15) prompted the US War Department to place a new order for 'common rifles' with the Harper's Ferry Armory. In the second production run (1814–18), 15,703 M1803 'short rifles' were delivered. Further production of M1803 rifles was organized, involving contractors such as Henry Deringer of Philadelphia, Pennsylvania, and Robert Johnson of Middletown, Connecticut, for 2,000 .54-calibre full-stock flintlock rifles at $17 apiece. The second production run contracts were signed with five private companies (Henry Deringer, Robert Johnson, Simeon North, Nathan Starr and R. & J.D. Johnson) in 1817 for the production of a similar rifle; the Harper's Ferry Armory was already focusing on the production of muskets. The M1817 'common rifle' had a .54-calibre barrel 36in (91.4cm) long with iron mountings, and a round brass patch box in the butt stock. Total output was 38,200 pieces up until 1840.

The .54-calibre 1807 Contract rifle had a 38in (96.5cm) barrel. (Photo courtesy of Morphy Auctions, www.morphyauctions.com)

US BREECH-LOADING FLINTLOCK RIFLES

The M1819 Hall rifle, the first breech-loading flintlock rifle to enter service, was invented by John Hancock Hall and patented by Hall and William Thornton on 21 May 1811 (US Patent 1516X). (Thornton, Superintendent of Patents, had nothing to do with Hall's invention, but simply refused to issue the patent until Hall shared the ownership with him, stating that he was working on the same invention.) The heart of the new rifle was a hinged block that opened by pressing the release catch located under the receiver. A spring raised the muzzle of the block. The chamber was charged with black powder, and a round ball was pressed into its mouth. The action was closed, the pan was primed and the rifle was ready to fire. According to the Hall and Thornton contract, the Superintendent of Patents would receive 50 per cent of the income from selling the patent rights to private individuals, while Hall would own the privilege to establish a factory to produce the rifle. Hall invested his family fortune in the factory, but his 'partner' was interested in quickly selling the concept to entrepreneurs. Conflicts were unavoidable, making Thornton the key opponent of the concept until 1820.

Hall's goal was to sell his products to the US Government. He delivered five rifles (each $40) and three smoothbore muskets (each $30) for trials between December 1813 and November 1814. The trials resulted in an immediate order for 200 rifles at $40 apiece, to be fulfilled by 1 April 1815. Lacking the necessary infrastructure to facilitate production, Hall rejected the order but worked on creating the production capacity required. As he wrote to Lieutenant-Colonel George Bomford in January 1816, his goal was interchangeability, so that even if 1,000 rifles were dismantled and their components piled up, rifles could be built from components chosen at random (Smith 1977: 191). Hall was hoping for an order for 1,000 rifles at a price of $40 apiece, so he could finance the modernization of his workshop in Portland, Oregon. Despite the favourable trials, Colonel Decius Wadsworth, Chief of Ordnance, was not convinced about the immediate adoption of Hall's weapon and placed an order for only an additional 100 rifles at a unit price of $25, disappointing the inventor. The rifles were shipped to Captain George Talcott at the arsenal in Charlestown, South Carolina, for further field trials. Although Talcott was an avowed opponent of breech-loading, after careful inspection and repeated trials he was convinced and in December 1817 wrote to Captain John Morton, Deputy Chief of Ordnance, in favour of the Hall rifle, predicting that it would supersede the common rifle (Smith 1977: 192–93).

On the edge of bankruptcy, Hall tried to use his network of family friends to proceed, but all they could do was to direct Hall and the concept of interchangeability to the Harper's Ferry Armory. He was invited by Secretary of War John C. Calhoun to supervise the production of his design. Hall accepted the position and the first four samples – two smoothbore muskets and two .54-calibre rifles – were completed by October 1818, and were put to trial again by James Stubblefield, Superintendent of the Harper's Ferry Armory and Armistead Beckham, Master Armorer. These weapons were then submitted for further trials in Washington, DC. The committee of officers fired 7,061 shots with the smoothbore

THIS PAGE & OPPOSITE This 1832-dated M1819 Hall rifle was built by Simeon North. (Rock Island Auction)

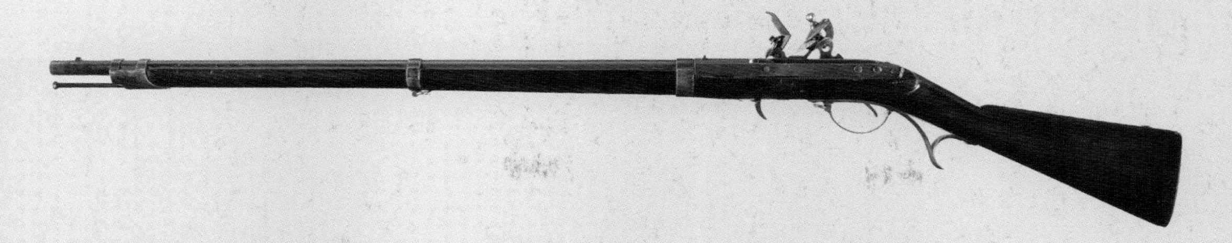

muskets and 7,186 shots with the rifles in a three-month period, proving the weapons' endurance and reliability. The committee's final report praised the rifle for its lightness, ease of use, superior accuracy and reduced recoil (Smith 1977: 195).

Hall was now hoping to receive a 10,000-unit order, but the committee only recommended the purchase of manufacturing rights for 1,000 rifles, with the option of making the guns in Hall's workshop at Portland, or at Harper's Ferry. Hall reluctantly accepted the offer on 19 March 1819 to receive $1 for each rifle and a monthly salary of $60 for modernizing and setting up the Armory's Rifle Works to produce his rifle. The first 1,000 rifles were delivered in 1824, and a second contract of 1,000 rifles was also approved by that summer.

The decision-makers were still sceptical, however, so Secretary of War James Barbour authorized a full-scale evaluation of the rifle in 1826. Two infantry companies at Fort Monroe, Virginia, were equipped with Hall firearms, one with standard smoothbore muskets and the other with .54-calibre Harper's Ferry-manufactured rifles. Both companies commenced a five-month course of practice in parallel. A detailed evaluation report was submitted to the Ordnance Department stating again the vast superiority of the rifle, thus supporting the interchangeability concept. In the meantime, a team of experts carefully examined every aspect of Hall's manufacturing process, including time, complexity, the price of machines and the cost of parts. The final committee report concluded that the production was capable of full interchangeability. As a result, Hall received his third contract, for 5,000 rifles (in 1828 increased to 9,000), and his salary was doubled (Smith 1977: 205–09).

Immediately after the investigations, 96 Congressmen requested samples of Hall's weapons. The cost of the rifles produced in the Rifle Works was covered by Federal funds, so these rifles could only be shipped to US troops. Manufacturing rifles to arm state troops would require the involvement of contractors, and Bomford recommended contracting Simeon North of Middletown, Connecticut. Hall opposed the involvement of another production facility, however, as neither the low cost he offered ($16.68 per rifle) or the exceptional quality of the Rifle Works could be replicated by a private entrepreneur, while a failure could harm the reputation of the breech-loading and interchangeability concepts. The Ordnance Department did not share Hall's fears and in 1828 contracted North to produce 5,000 rifles at $17.50 apiece.

In the event, Hall was right: North's machinery was inadequate, while the gauges supplied by the Ordnance Department were also faulty. The first deliveries arrived only in 1830, with many of the rifles failing inspection. Hall's production was in trouble as well. The US Government cut costs on the operation of the Harper's Ferry Armory, so important developments were cancelled, and many skilled craftsmen had to be laid off. The year 1834 was challenging again as Secretary of War Lewis Cass ordered the Rifle Works to become a branch of the Harper's Ferry Armory, demoting Hall to Master Armorer and tasking him with applying all points of the newly published *Ordnance Regulations* of 1834. Losing his independence was unacceptable to Hall, and he successfully employed all of his political contacts to counter this decision. He signed a new contract that granted him a director's position with an annual salary of $1,000 supplemented by $1,600 yearly for the right to use his machinery at Harper's Ferry and in any US arsenals (Smith 1977: 215–16).

Inspection of the Hall flintlock breech-loading rifle was a more complex task than with the common rifle. The barrel was proofed before the receiver was attached; the breech was closed with a proving plug and the barrel was loaded with the proof charge from the muzzle. These charges were the same as that used for the common rifle (OM 1841: 102–03). Parts interchangeability and function was checked as well, with inspectors taking apart any number of rifles, gauging the parts and putting them together from randomly selected components. 10 per cent of the first 400 rifles and 2.5 per cent of the following batches of 400 were to be selected for further examination (OM 1841: 112–13).

Hall's contract was renewed on an annual basis up until his death on 26 February 1841. The total production of Hall rifles at Harper's Ferry was over 19,000 Hall M1819 flintlock rifles and another 3,190 Hall M1841 percussion rifles (see page 33). Simeon North made 5,700 Hall flintlock rifles (Thomas 1997: II.133; Reilly 1972: 24). The cost of the contracted and Harper's Ferry-manufactured rifles did not differ significantly. The average cost of a Hall rifle was $16.32 in the 1830, while North received $17.50 per piece. Saving money was not a key goal of the Rifle Works, however; rather it was to carry out research into new production methods (Smith 1977: 220).

THE M1819 HALL RIFLE EXPOSED

.525-calibre Harper's Ferry M1819 Hall flintlock rifle

1. Buttplate screw
2. Buttstock
3. Flint
4. Frizzen
5. Rear sight
6. Barrel band
7. Front sight
8. Ramrod
9. Front sling mount
10. Bayonet
11. Cartridge in chamber
12. Catch
13. Catch spring
14. Main spring
15. Rear sling mount
16. Trigger guard
17. Trigger
18. Pistol grip
19. Axis pin
20. Cock screw
21. Frizzen screw
22. Breech tilted upwards
23. Cartridge powder tipped into chamber
24. Ball
25. Sear spring
26. Trigger stop pin
27. Axis pin hole
28. Tumbler
29. Cock
30. Touch hole
31. Frizzen spring
32. Frizzen spring screw
33. Cartridge powder in chamber
34. Ball

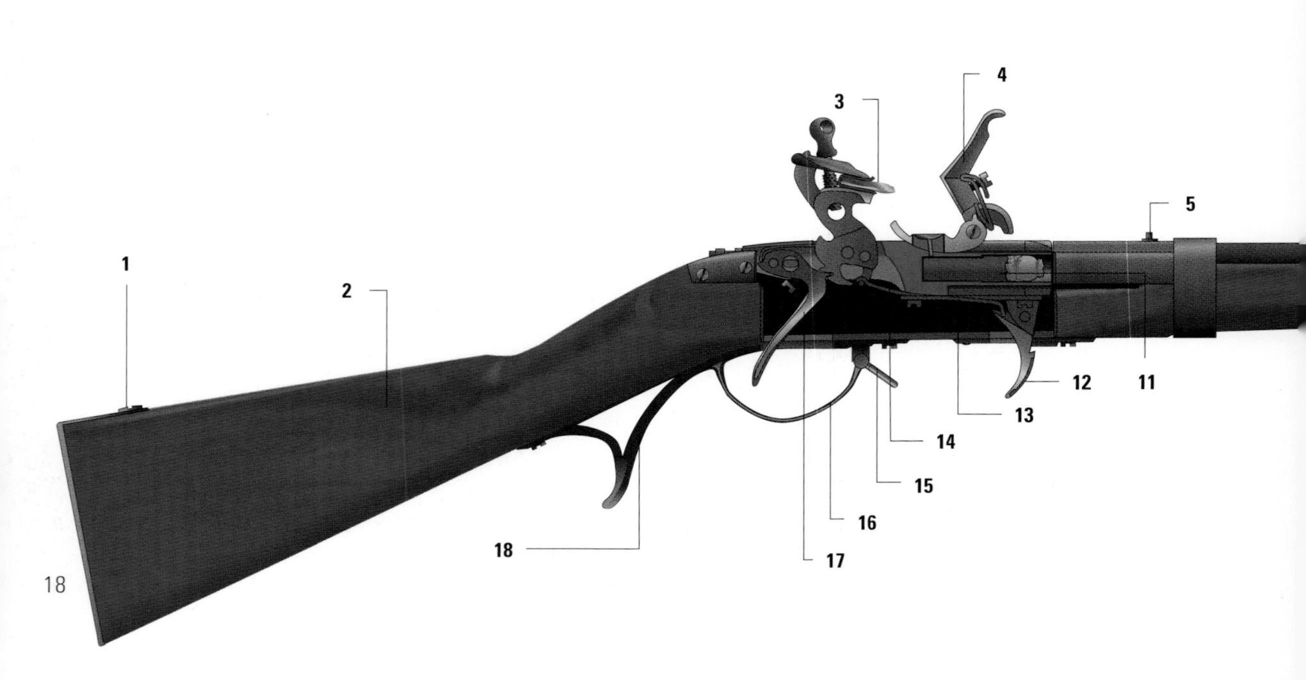

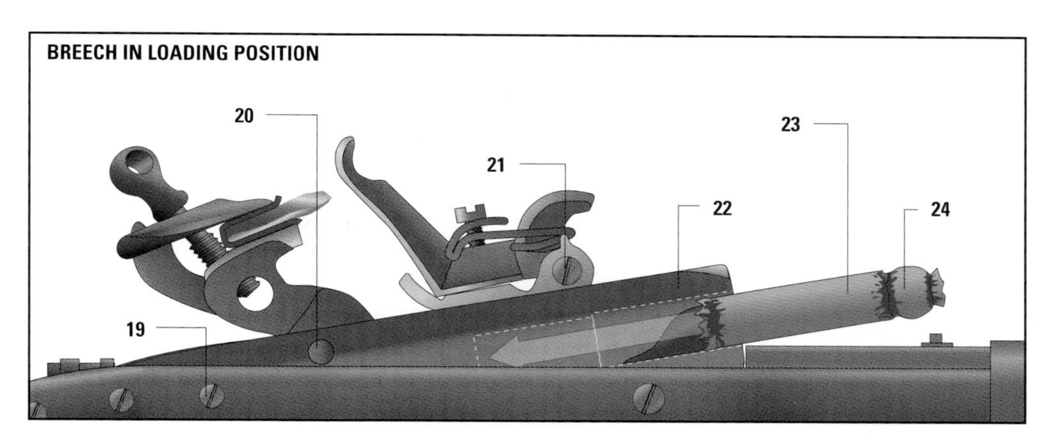
BREECH IN LOADING POSITION
20
21
22
23
24
19

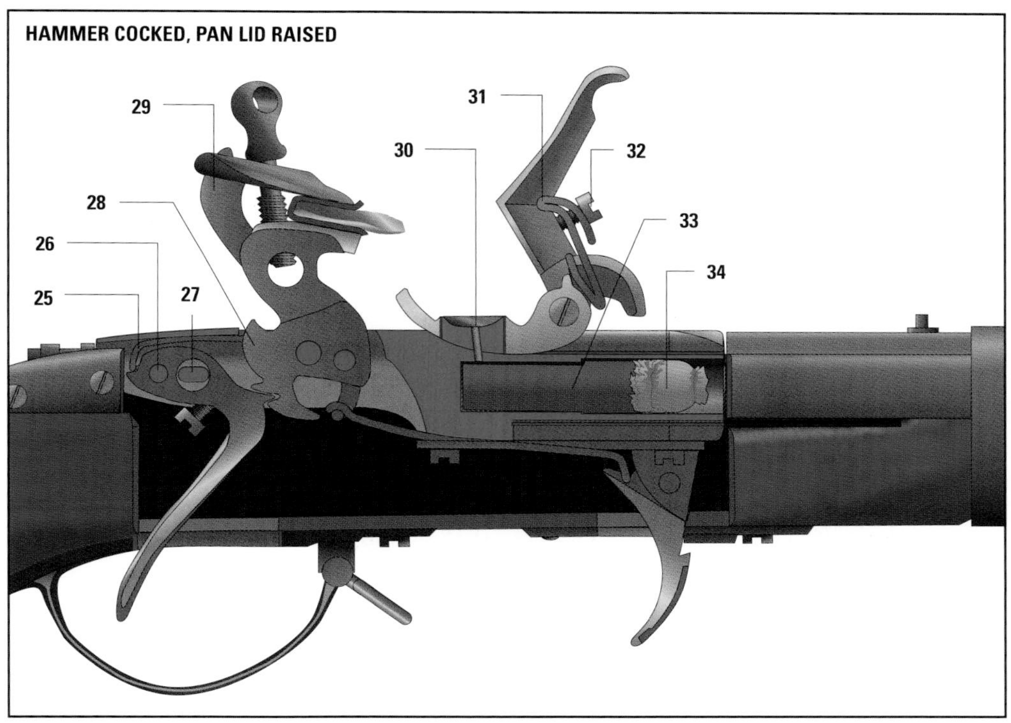
HAMMER COCKED, PAN LID RAISED
29
31
30
32
28
33
26
34
25
27

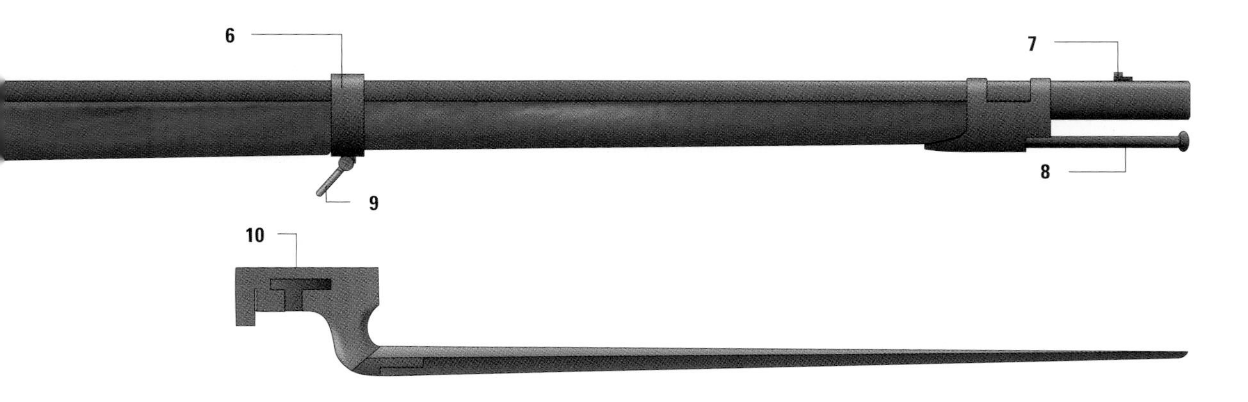
6
7
8
9
10

British flintlock rifles

The British Army raised its first rifle-armed troops in the mid-1740s, small units equipped with 'rifled carabines' sourced from Europe and fielded in North America. These weapons were most probably European *Jäger*-style rifles and were accompanied by their own bullet moulds. There is no indication that they were charged with finer powder than musket powder (Bailey 2002: 11–12).

A batch of 300 rifles was purchased in 1756, during the Seven Years' War, by Colonel Jacques Prevost, commander of the 62nd (Royal American) Regiment of Foot, a unit manned by German emigrants to North America. All of these rifles had iron rammers and were equipped with bayonet and bullet moulds. In 1761, a British light-infantry unit was raised in Europe by Major Simeon Fraser of the 24th Regiment of Foot, from 360 selected men of 12 infantry regiments. There is no particular proof that these soldiers were armed with rifles, but they were designated as '*chasseurs*' or '*Jägers*', and were attached to units consisting of light troops originating from Germany (Bailey 2002: 18). Although the British light infantry proved to be highly effective, all such units were disbanded after the Seven Years' War, and much of their hard-won knowledge and experience was lost.

At the outset of the American Revolutionary War, Colonel William Faucitt – a British Army officer involved in hiring German troops to serve in North America – advocated the issue of five rifles to each company of the Highlanders to increase the number of rifle-armed British troops available in North America (Bailey 2002: 22). George, The Viscount Townshend, Master-General of the Ordnance, obtained the Crown's approval to purchase 1,000 rifles based on two Prussian-pattern rifles supplied by Faucitt in December 1775. One rifle burst while proofing, but convinced by the performance of the surviving sample, Townshend placed an additional order for 200 pieces with a Hanoverian gunmaker, August Heinrich Huhnstock. These P1776 rifles were delivered in spring 1776, while the British Government contracted four Birmingham gunmakers to produce the remaining 800 pieces. William Grice & Son manufactured 600 of these rifles. The final batch of rifles produced by Matthias Barker, Samuel Galton & Son, William Grice & Son and Benjamin Willets was delivered in December 1776.

In the early part of the French Revolutionary Wars, the British Army continued to employ German riflemen, but the conflict provided the impetus to commence domestic rifle manufacture because a large portion of the French Army consisted of light troops, thereby increasing the British

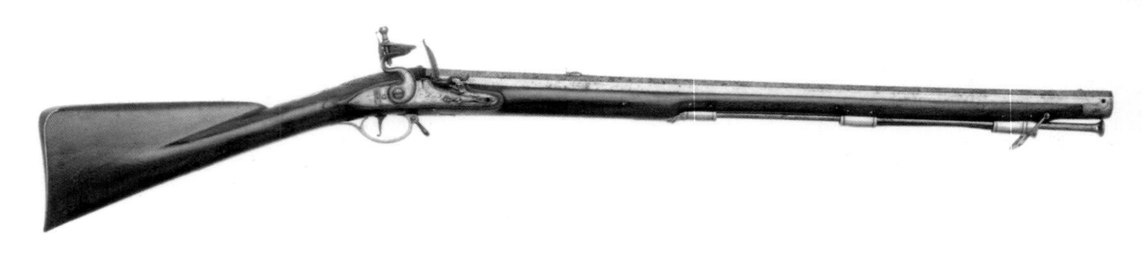

This John Hirst-made officer's rifle, a copy of the P1776 rifle, was made for a light-infantry officer in the 18th century. The British-made P1776 rifles were copies of the Hanoverian rifles. The pitch of the grooves was one turn in 27in (68.6cm) for the Hanoverian rifle, while the Birmingham-made copies had a twist-rate of one turn in 56in (142.2cm). Both rifles had a nominal carbine-calibre bore, but the land-to-land diameter of the Hanoverian bore was .620in, while it was .630in in the case of the British-built rifle. Both barrels were relieved at the muzzle to facilitate introducing the .615in ball into the bore (Bailey 2002: 24–25 & 198–99). (© Royal Armouries XII.5022)

BRITISH BREECH-LOADING FLINTLOCK RIFLES

The screw-plug breech-loading principle originated in the late 17th century, with Daniel Lagatz of Danzig. The idea was later copied by Isaac de la Chaumette, a Frenchman who even patented it in England, in 1721. The authorities did not pay much attention to policing patent infringements, however, thereby allowing this concept to reappear intermittently. The first screw-plug breech-loading rifles tested by the Board of Ordnance were submitted for trials in 1762 by John Hirst, a Tower Hill gunmaker. The trials were satisfactory and the Board of Ordnance ordered 20 pieces from Hirst in 1763. These rifles had a 36in (91.4cm) rifled round barrel, some with .65in carbine and others with .68in musket bores (Bailey 2002: 20).

Captain Patrick Ferguson of the 70th Regiment of Foot obtained a patent for his screw-plug breech-loading firearm on 2 December 1776. Ferguson's rifle was first mentioned in March 1776 in letters from Ferguson to Edward Harvey, Adjutant-General to the Forces. The first trials of Ferguson's rifles manufactured by London gunmaker Durs Egg were held in Woolwich on 27 April 1776. The inventor presented his rifle personally. An article in *The Gentlemen's Magazine* covered the occasion:

> Some experiments were tried at Woolwich before Lord Townshend, Lord Amherst, generals Harvey and Desagulier and a number of other officers with a rifle gun upon a new construction, by Capt. Ferguson of the 70th Regt.; when that gentleman under the disadvantages of a heavy rain and a high wind performed the following four things, none of which had ever before been accomplished with any other small arms. 1. He fired during 4 or 5 minutes at a target at 200 yards [183m] distance at a rate of 4 shots each minute. 2. He fired 6 shots in one minute. 3. He fired four times per minute advancing at the same time at a rate of 4 miles in the hour. 4. He poured a bottle of water into the pan and the barrel of the piece when loaded so as to wet every grain of the powder and in less than half a minute fired with her as well as ever without extracting the ball. He also hit the bulls-eye at 100 yards [91m], lying with his back on the ground and notwithstanding the unequalness of the wind and wetness of the weather, he only missed the target three times during the course of experiments. (Urban 1776: 283)

The rapidity of fire, accuracy and ease of cleaning a 'dead' charge (one that failed to ignite) from the breech all played an important part in shaping favourable opinion of the rifle. The dead-charge issue was especially persuasive, as this was a difficult process with muzzle-loaders, and was impossible during combat. Examples of Ferguson's rifle were presented to the King and Queen on 2 October 1776 in Windsor Forest. Ferguson trained six of his men for the trial, but the frightened soldiers did not perform well. Ferguson seized one of the rifles and fired three shots lying on his back and another six standing in less than two minutes, with excellent group (Roberts & Brown 2011). The cartridge held a .615in carbine ball, and three drams (5.316g, 82 grains) of 'superfine double-strength' (SDS) powder. Both paper cartridges and loose powder in horns and naked balls were issued to the troops, indicating the dual loading method of the rifle: paper cartridge for skirmishing, loose powder and ball for sharpshooting (Bailey 2002: 56–57). The horn had a threefold purpose: measuring the charge for paper cartridges, pouring powder into the breech and priming the lock.

As production of P1776 rifles had ceased, the Board of Ordnance ordered the immediate production of 100 Ferguson rifles for further trials. The contracted partners for these rifles – Matthias Barker, Samuel Galton & Son, William Grice & Son and Benjamin Willets – were paid £4.0.0 per rifle. All of the Ferguson rifles produced by the four contractors were inspected by Ferguson himself, and they were delivered in January 1777 (Bailey 2002: 39).

This Ferguson rifle was submitted to the Board of Ordnance by Durs Egg as a pattern for later production. Each production rifle's breech plug was made of brass so it would not rust into its place; moreover, brass cooled faster than steel to facilitate opening. The plugs were not interchangeable, each being numbered to one particular rifle. (Photo courtesy of Morphy Auctions, www.morphyauctions.com)

Army's need for regular light troops to counter them. The Board of Ordnance carried out trials for various rifled bores and rifles in Woolwich during January and February 1800. Eight barrels were tested, including one each from Durs Egg, Ezekiel Baker, Samuel Galton, Henry Nock and 'Mr German' plus three other barrels of French, German and American origin. Each barrel was first shot to 300yd (274m) from a fixed mortar bed, then remounted into the stock and shot from the shoulder at the same 11ft×9ft (3.35×2.74m) target. Baker's carbine-calibre rifle – with a barrel length of 30in (76cm) – was only test-fired from the mortar bed but performed extremely well, hitting the target 11 times out of 12 shots (Baker 1806: 23).

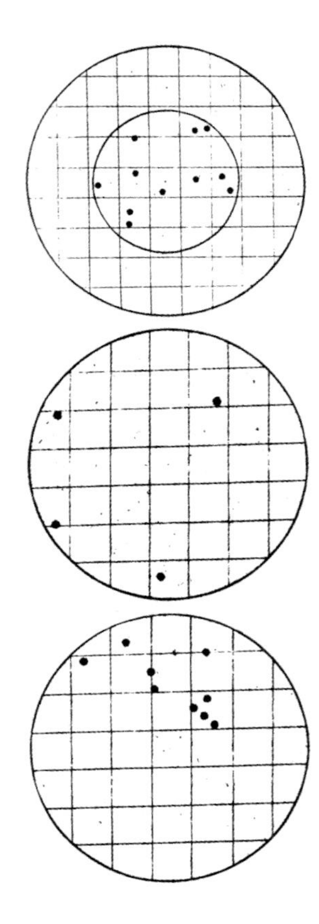

These illustrations from Baker 1806 show (above) the group of the Baker rifle on a 9ft (2.74m) target at the official Board of Ordnance trials on 4 February 1800 and (below) a comparison of the performance of two rifles, one made by Henry Nock and the other by Ezekiel Baker, at trials held for HRH The Prince of Wales on 4 June 1803, showing both rifles' groups on a 7ft (2.13m) target at 200yd. (Author's collection)

The Baker rifle's slow twist-rate was justified by the maker with a series of experiments: 'I cut it to one foot [30.38cm] one quarter turn, and found I could fire more true at short distance than I could when more angle in rifle. From this I made a barrel 2 feet 6 inches [66cm], and rifled it one quarter turn, and found I could range farther and more true than ever with any before, and with less elevation.' (Baker 1806: 22). Reducing the angle of the rifling was a good response to the increased powder charge of the rifles – the SDS powders accelerated the ball to a velocity at which the rifling lost its grip on the soft lead ball, destroying the rifle's accuracy. The slower twist-rate was much more suitable for high muzzle velocities. All of the Baker rifle's competitors had a faster twist-rate of ½–1 complete turn.

Baker submitted the pattern rifles to the Board of Ordnance on 12 March 1800, and these were used to make further pattern guns. The Board of Ordnance contracted private companies to manufacture parts of the rifle. The best London and Birmingham gunmakers supplied barrels, stocks, iron and brass furniture, locks, tools and sword-bayonets. The first batch of 1,013 rifles was registered into the Board of Ordnance stores in September 1800. The first order covered 1,500 iron barrels forged from skelps (narrow strips of rolled or forged metal), and 1,500 Damascus barrels forged from twisted iron. The greatest bottleneck that hindered early production was the rifling, as only Durs Egg, Henry Nock and Ezekiel Baker possessed rifling benches (Bailey 2002: 108). The 30.25in (76.8cm) barrel had a hooked breech plug, fitting into the false breech screwed to the stock. The walnut stock resembled the full-length *Jäger* stocks. Not all of the rifles of the first batches had a patch box, but each had a cheek piece to offer a comfortable hold. The iron ramrod had a flared tip with a perpendicular hole for inserting a torque bar for pulling the ball.

The total production of Baker rifles to 1815 amounted to 22,339 pieces: of these, 4,982 were Tower made, 4,192 were manufactured by the London trade and 13,164 originated in Birmingham (Bailey 2002: 119). The Royal Manufactory Enfield did not commence operations until 1816, and its manufacturing capacities only became essential when the Board of Ordnance approved the order of 10,000 new Baker rifles. In 1823 it was decided to split the production of the new rifles between the Royal Manufactory Enfield and the London trade, making use of parts from the Birmingham trade. Production continued until 1827. The grand total for

A P1800 Baker rifle with P1801 (second type) sword-bayonet. The first-type square knuckle-guard sword-bayonet had a 58.1cm-long blade and attached to the rail on the right side of the muzzle. The square knuckle-guard was changed to a 'D' shape on the P1801. When the troops requested socket bayonets like those used by line infantry, the Board of Ordnance decided to convert existing P1800 Baker rifles by shortening the stock at the muzzle, replacing the front sight, fitting a new nose cap and removing the bayonet rail. The resulting P1800/15 rifle was only a temporary solution, and although the 450 P1820 Baker rifles were ordered with this bayonet, in 1823 the Board of Ordnance reverted to the old method of mounting. (© Royal Armouries XII.148 & X.132)

the new rifle was 11,432 pieces. These rifles had a round lock plate and a swan-neck hammer. The bayonet attached to the muzzle with the rail again; and the blade was a combination of the old sword type and a triangular blade. The last production batch of the Baker rifle was 1,013 pieces in 1838–39 (Bailey 2002: 137–39).

RIFLES OF THE PERCUSSION ERA

Military rifle development between 1815 and 1850 focused on three aspects: replacing the flintlock with a more efficient form of ignition; simplifying the loading procedure to match the firing rate of the smoothbore musket; and reducing the calibre of the bore to have a lighter cartridge and flatter trajectory. During this period the French would join the Prussians, Austrians, British and Americans in pushing forward the limits of rifle technology.

The flintlock system had serious drawbacks. The security of ignition was greatly dependent on many factors: the sharpness of the flint, the cleanness and hardness of the frizzen, the obstruction or clearness of the touch hole, and finally humidity. The musket could misfire easily as the flint became blunt, or if the touch hole was obstructed. Out of ten shots, usually two and three misfired. The percussion system would resolve these problems by offering a fast and secure ignition, reducing the rate of misfires to less than 1 per cent. The heart of the system was the percussion compound, which exploded upon impact. The percussion system was derived from Scottish clergyman Alexander John Forsyth's invention, patented in 1807, but the percussion cap itself would be a result of parallel inventions.

Flash in the pan: the moment of flintlock ignition. Upon pulling the trigger the flint in the hammer jaws fell on the frizzen, shearing off hot metal particles from the metal surface while opening the pan. The sparks fell on the priming powder and flames ignited the main charge in the barrel through the touch hole. Although this happened in fractions of a second, the lock time was so slow that the soldier could easily pull the weapon off-target while the ball was still in the barrel. The flash in the pan disturbed the aiming eye, while the hot gases blowing out of the touch hole disturbed the neighbouring soldier in the line. (Author's collection)

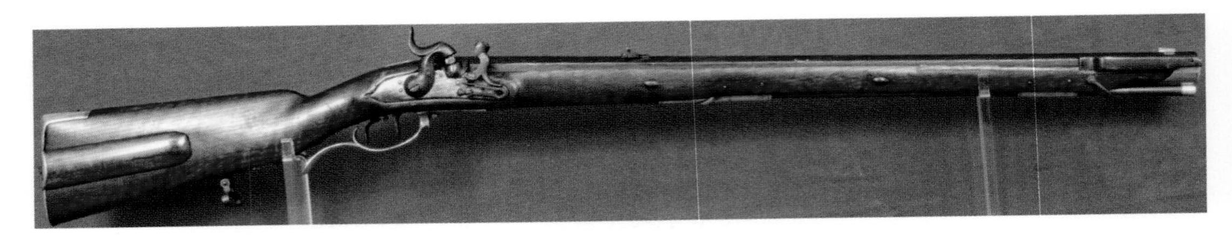

This Potsdam-made M1810 *Neue Corpsbüchse* has been converted to percussion. The conversion was done by welding a drum to the right side of the breech, adding a nipple, replacing the flintlock hammer with a percussion one, removing the pan and replacing the frizzen with a safety device that held the hammer away from the percussion cap when engaged. Note the imperfect fit of the safety device. (College Hill Arsenal)

Prussian percussion rifles

The first Prussian firearms to be converted to percussion were the M1810 rifles, during 1831–34. Conversion was not trouble-free, however, as the lack of proper military standards in the manufacturing and the great variation of the rifles in service required intensive hand fitting. The Ministry of War therefore decided to replace the flintlock rifles entirely with a rifle based on the original specifications of the first Potsdam-made rifle from 1810, to be called the M1835 *Jägerbüchse*. Although the calibre remained the same, 14.64mm, the rifling profile was changed. A slower twist-rate of 1:93cm was applied to increase accuracy as the percussion ignition raised the pressure of the charge and increased muzzle velocity. The welded breech was also replaced with a modern-patent breech, resulting in a fast and secure ignition. Many of the older rifles' barrels were also replaced with these new ones, but as the number of rifle-armed troops in the Prussian Army was substantially reduced in 1840, leaving only the Garde-Jäger-Bataillon and Garde-Schützen-Bataillon in service, production numbers of the new type were limited.

The next rifle to enter Prussian service was the revolutionary Dreyse *Zündnadelgewehr* ('needle-rifle'). Having learned about percussion ignition while working for the gunsmith Jean Samuel Pauly in Paris during 1809–13, Johannes Nikolaus Dreyse worked on developing a breech-loading design; it was declared to be classified material by the Prussian Ministry of War in 1831. Dreyse successfully presented his breech-loading rifle firing a self-contained cartridge to the Ministry in 1836. The rifle fired five shots per minute, and all hit the target. Dreyse fired the last seven shots after contaminating the action with a handful of dirt; the rifle still operated, and hit the target seven times. The Ministry was fully convinced and ordered 150 rifles for field trials. Dreyse received 90,000 thalers for building a factory in Sömmerda to produce 60,000 rifles, and production of the M1841 *Zündnadelgewehr* began on 15 October 1841. The rifle was still classified material, so the codename of the new firearm was *leichtes Perkussionsgewehr* ('light percussion rifle').

The first Prussian firearm to fire an elongated bullet, the M1835 *Jägerbüchse* was intended to replace all of the M1810 rifles in service. It was based on the design of the original M1810 rifles made in Potsdam, and soldiered on until the end of the 1850s. The Ministry of War selected the Thouvenin pillar-breech system (see page 67) and installed a 7mm-thick and 40mm-long pin into the patent breech. The rifle fired a 14.38mm-diameter bullet weighing 31g that had to be 'upset' on the pillar with the heavy thrusts of the iron ramrod (Götz 1978: 167). (© Royal Armouries XII.2289)

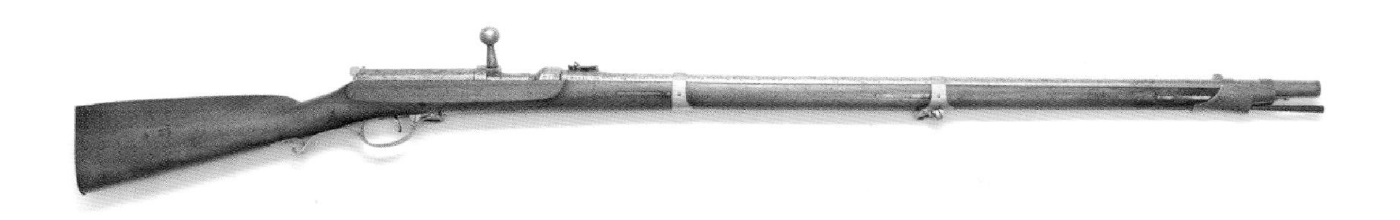

The revolutionary M1841 Dreyse *Zündnadelgewehr* ('needle-rifle') fired a self-contained cartridge. The rifle was 142.5cm long overall with a 90cm-long barrel. The twist-rate of the 15.43mm-calibre bore was one complete turn in 74cm, quite fast for round balls. It fired a paper cartridge uniting the undersized bullet, the carton wad, the powder charge and the priming compound. The first Dreyse projectile was a round ball, but in 1847 it was replaced by the *Bolzengeschoß*, an early cylindro-conical bullet inspired by the work of Capitaine Henri-Gustave Delvigne and Major Louis-Étienne de Thouvenin (see page 67). The bullet calibre was smaller than that of the bore, and it was inserted into the carton wad, or *Spiegel* that created the interaction between the rifling and the projectile. The priming compound was placed on the bottom of the wad. Upon firing, the long steel pin moved forward, penetrated the powder charge and exploded the primer. (Author's collection)

Habsburg percussion rifles

The Habsburg Army was aware of the advantages of the percussion system, but rearming the entire Army was a formidable challenge as the financial resources of the Habsburg Empire were exhausted. The armed forces had 600,000–700,000 flintlock firearms on hand, an enormous asset that had to be saved. It was also believed that it would be very difficult to retrain the soldiers to follow a completely new loading process; they needed a percussion-loading procedure close to that used for the flintlocks. Natalis-Felix Beroaldo-Bianchini, head of the Vienna Arsenal, tested the new percussion-cap system in 1829 and deemed it unfit for military service as the caps were too small for the soldiers to handle. Beroaldo-Bianchini was right: the small (civilian-gauge) caps he tested were hard to handle; larger, winged caps were also available, though, but were not tested (Gabriel 1990: 71–72).

A customs officer from Milan, Giuseppe Console, patented a tube ignition system in 1831. The principle was simple: he loaded black powder mixed with potassium-chlorate into a piece of straw. It exploded upon impact, resulting in a fast ignition. Only two parts of the lock had to be changed – the pan and the frizzen – and a steel block had to be put into the jaws of the hammer. The new pan accepted the priming tube. The cover (formerly the frizzen) was shut on the priming tube. Upon firing, the hammer fell on the cover and a sharp edge exploded the tube (Gabriel 1990: 72). The solution seemed cheap and easy, and the loading method was similar to that of the flintlocks.

The first trials of the new ignition system, conducted by the 6th Feldjäger-Bataillon, began at Eger in the spring of 1835. The converted M1807/35 *Jägerstutzen* rifles performed well. On 26 August 1835, Emperor Ferdinand I (r. 1835–48) nominated a committee of three officers to examine the possibility of rearming the entire Habsburg Army with Console's invention. One member of this committee was Vincenz von Augustin, principal inspector of handguns and artillery of the Arsenal (Götz 1978: 144). A Console-armed soldier fired 12 shots with the new rifle; during the same time period a flintlock-armed soldier was only capable of firing five shots (KA, HKR 1835 D36/21). The new rifles were more accurate, and the new priming system proved to be less sensitive to the elements. On 9 January 1836, Ferdinand I ordered all *Jäger* battalions to be equipped with the new system (KA, HKR 1836 D36/1).

Beroaldo-Bianchini, a noted opponent of the percussion system, was not happy with the new system, however, and raised some valid concerns (Gabriel 1990: 74–75). The percussion tubes were attached to the paper cartridge with a piece of wire, which could foul easily in the cartridge box.

Giuseppe Console invented a percussion-tube mechanism for the conversion of infantry firearms. The first firearms to be converted as a test run were M1807 *Jägerstutzen* rifles. Even if the percussion ignition had great benefits compared to the flintlock, however, the Console lock was not ideal for military service. (Museum of Military History, Budapest)

If the pan cover was shut too quickly it could ignite the percussion tube. The wire of the percussion tube was too long, so if the tube accidentally fell between two balls in the cartridge box, the cartridges could go off. Moreover, the explosion of the percussion tube was too close to the face of the soldier and particles could fly into the aiming eye.

Vincenz von Augustin took over the modernization project, his greatest challenge being the lack of interchangeability. Console's lock needed a very accurate fitting: if the pan and the flash hole were not in line, the ignition was insecure. Augustin tried to overcome the problems of interchangeability and to meet Beroaldo-Bianchini's requirements, but failed. He had to redesign the lock completely, thus losing the opportunity to convert flintlocks cost-effectively. Augustin's modification was accepted for service on 21 December 1840 (KA, HKR 1840 D1/47). He managed to achieve full interchangeability regarding the locks, but the old rifle and

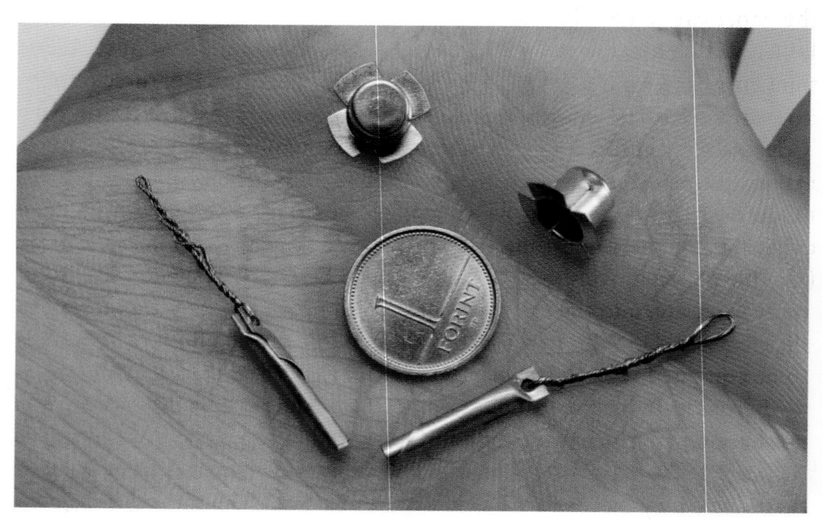

Here, reproduction percussion tubes as improved by Vincenz von Augustin are shown alongside large four-winged percussion caps (larger than the type tested by Beroaldo-Bianchini), with a 16.3mm-diameter coin for scale. Console's straw priming tube had been a huge leap forward compared to flintlock ignition, but was still far from the reliability offered by the caplock system, due to the poor fitting of the lock to the barrel. The misfire ratio of the Console lock was around 10 per cent; Augustin managed to reduce it to 6 per cent – better, but far from the desired level of less than 1 per cent (Dolleczek 1970: 86). (Author's collection)

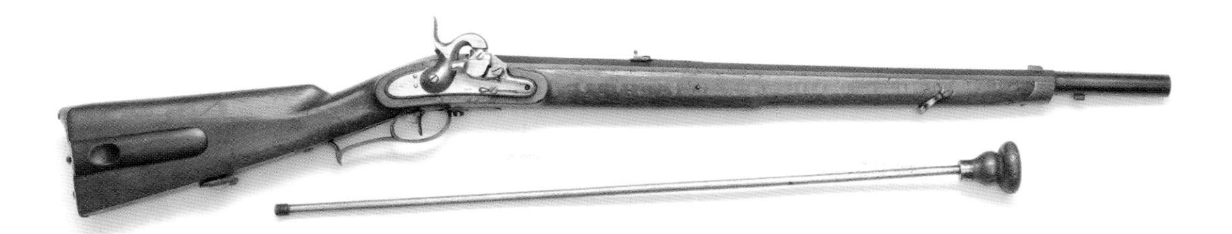

An M1842 *Jägerstutzen* rifle. Augustin perfected the system by completely redesigning the lock. This was the first lock system of the Habsburg Empire to reach full interchangeability. (Author's collection)

musket stocks had to be disposed of; even if the old barrels could be reused, they would have needed significant and costly modifications.

During the 1830s the Habsburg Army decided to increase the proportion of rifle-armed personnel among its troops, giving rifles to the first- and second-rank troops of the *Jäger* battalions by replacing the M1809 *Jäger* smoothbore carbines with rifles. The bores of 47,000 muskets captured from the French during the Napoleonic Wars were shortened from 84.7cm to 79.3cm and rifled with 12 deep grooves making a half-turn in the barrel. Each rifle received an Augustin tube lock, and a patch box was added to the stock. After the first successful field tests Ferdinand I ordered all of the *Jäger* battalions to be equipped with the new rifle, known as the M1842 *Kammerbüchse*, on 4 July 1844. Further endurance trials were held in the Vienna Arsenal until April 1845, involving two rifles that were shot more than 20,000 times without exhibiting any major defects. All of the *Jäger* carbines were replaced before 1846 (Götz 1978: 211).

In 1842, Augustin had made improvements to the barrel breech system of a French Army officer, Capitaine Henri-Gustave Delvigne (see page 34). The original chamber had sheer walls; Augustin rounded the sharp shoulders of the chamber to achieve a more even expansion of the hammered bore. The M1842 *Kammerbüchse* was one of the very first European rifles to utilize this solution. The chamber was useful for setting up the ball, but also to prevent the powder from being crushed, thereby aiding accuracy. The original M1842 *Kammerbüchse* cartridge held a 17.74mm-diameter round ball wrapped in a linen patch tied to the end of the paper cartridge. The bore had a nominal diameter of 18.1mm, so the bullet fitted tightly. The cartridge held a patched ball which was not as tight-fitting as the bullets of the contemporary *Jägerstutzen* rifle, but was still not as easy to load as the loose-fitting balls of a smoothbore musket. The powder charge was 55 Gran (3.85g, 59.4 grains) of *Scheibenpulver*, resulting in a muzzle velocity much lower than that of the smoothbore musket, so fitting a proper leaf rear sight was inevitable. The rear sight consisted of a block sight and two leaves, serving from 100 to 400 paces (75–300m) (Németh et al. 2012: 60).

The M1842 *Kammerbüchse* was designed for the first and second ranks of the Habsburg Army's *Jäger* battalions. The rifled bore calibre was larger than that of the infantry musket. The heavy ball required substantial elevation for accuracy at long range, so a leaf sight was attached to the bore. (Author's collection)

THE *KAMMERBÜCHSE* REVEALED

18.1mm M1842 *Kammerbüchse*

1. Buttplate screw
2. Buttstock
3. Pan lid
4. Tooth
5. Rear sight
6. Barrel band
7. Barrel band spring
8. Barrel
9. Front sight
10. Ramrod
11. Bayonet catch
12. Front sling-swivel
13. Cartridge
14. Trigger guard
15. Trigger
16. Rear sling-swivel
17. Sear spring
18. Bridle
19. Pan
20. *Kern*
21. Touch hole
22. Pan lid raised
23. Percussion tube inserted into *Kern*
24. Main spring
25. Tumbler
26. Sear
27. Lock plate
28. Hammer strikes tooth
29. Tooth detonates percussion tube
30. Powder in breech chamber ignites
31. Breech plug
32. Lid spring

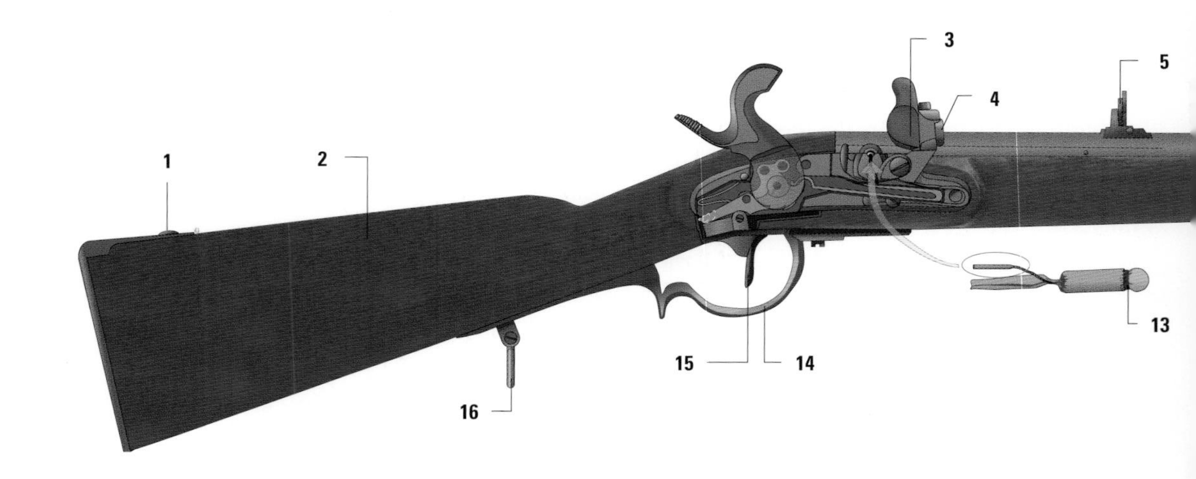

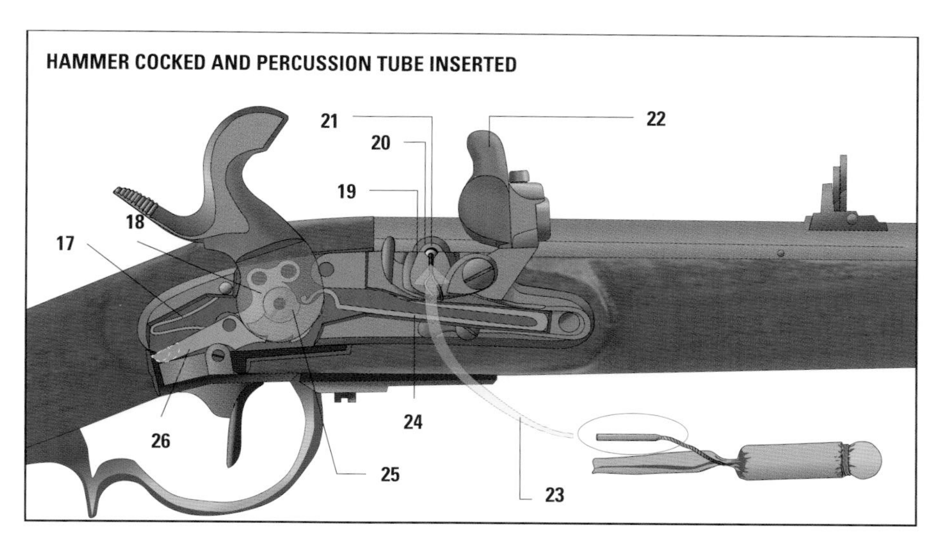
HAMMER COCKED AND PERCUSSION TUBE INSERTED
21
20
19
22
18
17
26
24
25
23

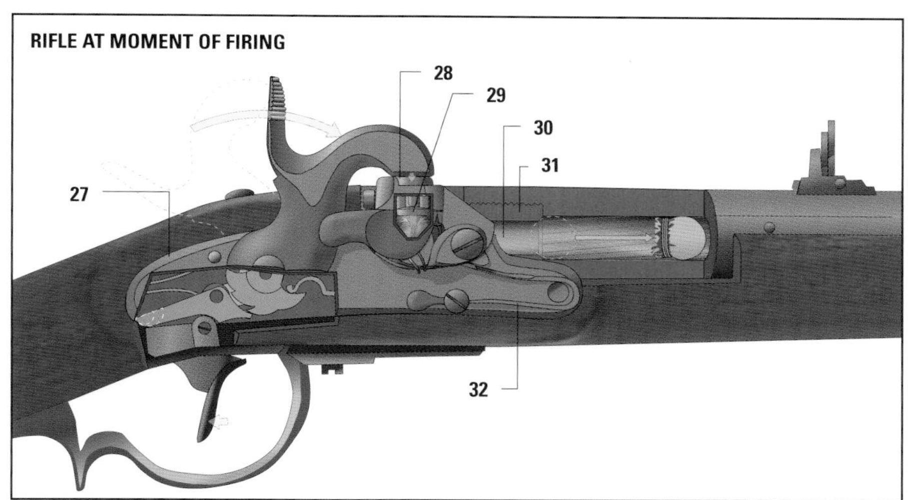
RIFLE AT MOMENT OF FIRING
28
29
30
31
27
32

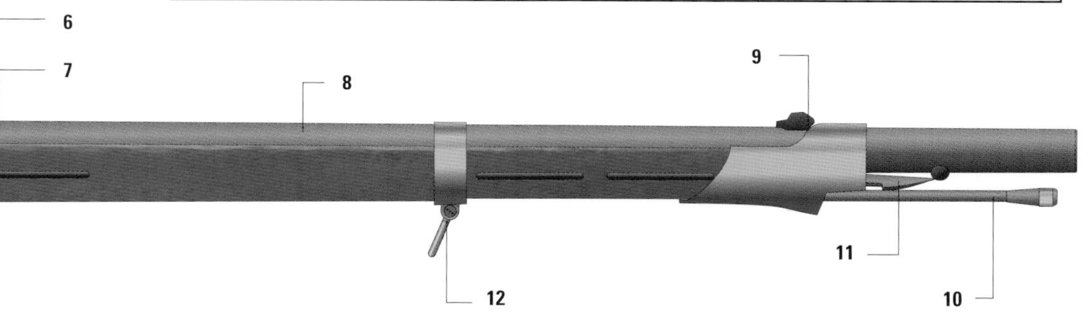
6
7
8
9
11
12
10

ABOVE & BELOW
Designed in the early 1830s to replace the Baker rifle, the .704-calibre Brunswick percussion rifle fired a patched belted ball from a two-groove rifled bore. Berner's design can be classified as a rifle-musket, having a 100.5cm barrel length. The calibre of the bore was 15.7–15.95mm and it had a two-groove rifling. The rate of twist was ¾ turn in the barrel, or one complete turn in 133cm. The 0.52mm groove was 13.8mm wide at the breech and narrowed towards the muzzle, reaching 7.3mm. In addition, the Brunswick was equipped with front and rear sights and a 40cm-long socket bayonet. The two-groove rifling aroused great interest all over Europe, being tested in Hanover, Prussia, Austria and Britain. Although the system seemed like an effective solution that combined accuracy with a high rate of fire, the rifle was extremely sensitive to fouling, rendering it hard to load with *Passkugel* ammunition after only 14 shots. Also, the ballistics of the mechanically fitting heavy belted ball were inferior compared to those of the Baker rifle. (Author's collection)

British percussion rifles

The Baker rifle's replacement in British service, the Brunswick, had its origins in the German state of that name. In 1831 Hauptmann (later Major) Carl Wilhelm Ernst Berner, a Brunswick veteran of the Peninsular War (1807–14), recommended an improvement of the seven-groove military flintlock rifle then in Brunswick service. Believing that the whole infantry needed a firearm that was as easy to load as a shotgun and as accurate as a rifle, Berner advocated reintroducing the tight-fitting *Passkugel* and easy-loading *Rollkugel* cartridges for the existing rifles, replacing the flintlock mechanism with a percussion lock and introducing his own development, the 'oval' bore. The Duke of Brunswick ordered a first batch of rifles in 1835. The rifle-musket fired two type of cartridges: the *Passkugel*, with tight-fitting unpatched balls and 4.4g (67.9 grains) of gunpowder; and the *Rollkugel*, with 5.5g (84.9 grains) of gunpowder. These cartridges also held a greased tow wad for substituting the lubricating function of the patch and the lead round ball.

Berner's system was adopted by smaller German states including Oldenburg, Hamburg, Bremen and Lübeck, but the two most important military powers to equip their rifle troops with two-groove rifles firing belted balls were Britain and Russia (Götz 1978: 197–204). George Lovell of the Royal Manufactory Enfield was the Brunswick rifle's principal advocate in Britain. During the early 1830s Lovell had worked on a percussion rifle-design featuring a traditional 11-groove rifling but with faster twist than the Baker's. His rifle was fitted with a new lock design, the back-action lock: the main spring was placed behind the hammer, leaving more wood near the breech area and rendering it stronger in the most critical part. Lovell was ready with his rifle in 1836 when the first two-groove rifles arrived from Brunswick. Lovell was the first to test the Brunswick rifle, and it proved to be superior to his own design. Although the trials did not convince the entire Board of Ordnance, the decision was made to adopt the rifle with a calibre of .704in and barrel length of 33.06in (84cm).

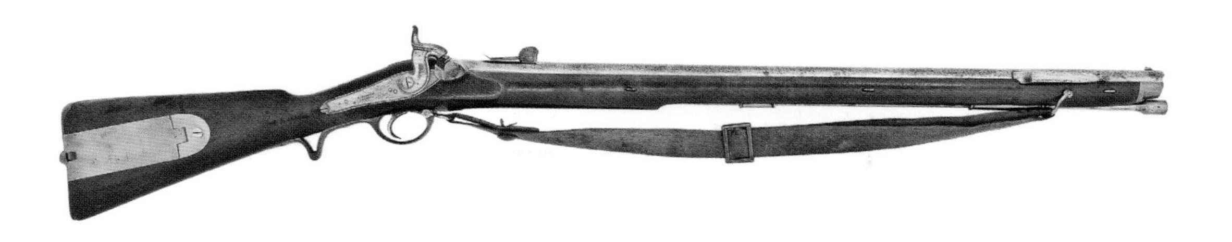

The Imperial Russian Army raised its first rifle regiments, armed with German-style rifles, in the last decades of the 18th century. Featuring a long-range sight, the .72-calibre M1825 Tula flintlock military rifle adopted in the 1820s had a six-groove barrel and an overall length of 120.6cm. Like the British, however, the Russians decided to adopt a percussion rifle based on the two-groove Brunswick rifle. Weighing 4.34kg and measuring 124.2cm overall, the M1843 Lüttich (Liége) carbine or *Littichskij Stutzer* fired a 33g patched belted ball; muzzle velocity was 337m/sec, which was lower than that of the smaller-calibre military rifles of the period and resulted in a more curved trajectory. The sights could be set up to 1,200 paces (852m). In 1848 a new conical bullet design, the Kulikov bullet, was accepted for the M1843; this ogival projectile had two protrusions close to the bottom to facilitate the required mechanical fit in the grooves. The Kulikov bullet offered better performance than the patched belted ball, but the projectile's increased weight resulted in a lower muzzle velocity and a more curved trajectory. Intended for the rifle regiments and the skirmishers of the Black Sea Cossacks, a total of 20,756 M1843 Lüttich carbines were issued to members of the Imperial Russian Army in 1849; a very low quantity in the million-strong Army (Mavrodin & Mavrodin 1984: 21). (Private Collection Photo © Christie's Images/Bridgeman Images)

The .704in-calibre Brunswick rifle was just as accurate as the Baker rifle at close range and proved superior at longer ranges, but was considered to be too heavy. The belted ball was also too heavy, while the powder charge was limited, resulting in low muzzle velocity and a curved trajectory. The inferior ballistics required more complicated sights and more effort during training. In November 1836 Lovell submitted six pattern rifles to the Board of Ordnance with some important modifications. First, he reduced the calibre to .654in; second, he increased the twist-rate of the rifling to one complete turn in the barrel; third, he introduced two semicircular cuts in the muzzle to aid the proper fitting of the patched belted ball. An immediate order for 1,000 rifles was placed, but as the Royal Manufactory Enfield was unable to produce the whole quantity in time, private contractors were involved as well. In 1837 the Board of Ordnance agreed to new modifications and an additional order of 2,000 rifles. The calibre of the rifle was changed back to .704in to simplify logistics.

US percussion rifles

The history of the cap in the United States is related to Joshua Shaw, who patented his copper percussion cap on 19 June 1822 and offered his invention to the US Government. Lieutenant Colonel George Bomford immediately modified a Hall breech-loader to Shaw's patent ignition: '... one of the patent rifles that is loaded at the breech has been fitted up for the application of your detonating powder, and the few trials which have been made with it have proved very satisfactory' (Report No. 1375). At the request of the Ordnance Department Shaw designed an enlarged version of the percussion cap to be used on artillery pieces, which he patented in 1828. Shaw claimed that his invention resulted in more certain, rapid and effective ignition; it facilitated the loading procedure of all firearms while requiring less powder in the main charge; and it was safe against accidental explosion. The US Government did not buy the patent, but the Ordnance Department set up the production of caps in the Federal arsenals and involved private contractors without proper agreement with Shaw. The inventor had to wait until 30 December 1844 for Secretary of War William Wilkins to authorize a payment of $18,000 to him (Report No. 1375).

Designed for the US Army's newly raised Dragoon regiments, the M1833 North-Hall and Harper's Ferry-made M1836 Hall breech-loading smoothbore percussion carbines were the world's first military

Percussion target or picket rifle in .424 calibre made by R.M. Wilder of Coldwater, Michigan. The third and final period of the long rifle was again a time of transitions. The acceptance of percussion ignition took off slowly in the United States, as on the frontiers it was much harder to acquire caps, while the flint was always available. The development of the percussion lock had an effect on the design of such rifles, while many old rifles were converted. The better availability of good fine powders and lead allowed a reduction in the barrel length and enlargement of the calibre. The production methods changed as well. The establishment of factories such as the Henry factory in Boulton and the Leman Rifle Works in Lancaster, both in Pennsylvania, industrialized rifle making and forced many individual gunsmiths to quit the business or to work for the new ventures (Kauffman 2005: 24–25). (Author's collection)

firearms with the new ignition. On 3 February 1837 the US Senate ordered the Secretary of War to carry out intensive trials of the new breech-loading small firearms invented by Hall, Cochran, Colt and Hackett, comparing their performance to the service smoothbore musket. The board of officers submitted their report on 15 May 1837. They examined the rate of fire, recoil, penetration, heating of the barrel in rapid firing, capacity of being used as a rifle, cost and simplicity of construction, durability, saving on ammunition, number of cartridges to be carried by a soldier and the advantages when used by and against cavalry. The report proved the superiority of the Hall breech-loader. It fired five shots per minute while the standard musket was only capable of three. Recoil was 'not sufficient in any of them to produce any great inconvenience in service ...' (Report 1837: 15). The penetration of the bullets fired by the breech-loaders was also equal or better than that of the large-calibre musket. The Hall breech-loader charged with 110 grains (7.13g) of rifle powder penetrated 2.175in (5.5cm) of seasoned white oak, while the regulation musket penetrated 2.3in (5.8cm) with a charge of 135 grains (8.75g) of rifle powder. The heating of the barrel was also favourable: 'Guns that receive their charges at the breech have been found to acquire a less degree of heat from a continuous course of discharges than the common arm, from the circumstance of their admitting, during the loading process, a free circulation of air through the barrels' (Report 1837: 15). Suitability for light-troop duty showed the clear superiority of two of the breech-loaders: '... it is consequently believed that Hall's rifle and Hackett's gun are the only descriptions of arms loading at the breech that can be considered as suitable for the service of light troops ...'. The simplicity and cost of construction also proved to be of benefit: 'The guns of Hall, and apparently of Hackett (...) are more easily managed, present less accountability to the soldier, are less liable to get out of order, and when they are so are more easily repaired' (Report 1837: 15). Cost-wise, Hall's breech-loader was the winner – probably not surprising as it was Hall who inspected the cost of cartridges, accessories and manufacturing. The clear winner of the trials was Hall's system, but the Hackett breech-loader's performance also raised interest. Cochran's percussion repeater was considered to be dangerous and unfit for service, while Colt's percussion repeater could be utilized in special cases with necessary improvements. Sceptics still opposed the repeating system. According to the board of officers, the rapid fire of the repeating action would

This M1841 'Mississippi' rifle is dated 1850 and marked 'E. WHITNEY'. The Mexican–American War (1846–48) increased the US need for rifles, and production of the M1841 started at Harper's Ferry in 1846, with an output of 700 rifles in the first year; a total of 25,296 M1841s were made in the Rifle Works until the adoption of the M1855 rifle-musket (Smith 1977: Table I). Additional contracts were signed with private contractors: John Griffith of Cincinnati, Ohio, an Englishman and former employee of Hall at the Harper's Ferry Armory, contracted to deliver 5,000 rifles in 1842, but failed to fulfil his obligations. The contract was taken over by Remington of Ilion, New York. The next contract was signed in 1845 with Robbins, Kendall & Lawrence of Windsor, Vermont for 10,000 rifles at $11.90 each, followed in 1848 by a second order for 15,000 rifles at a price of $12.875 each. (This order was supplied under the name Robbins & Lawrence, as Nicanor Kendall sold his share of the company to his partners.) Edward K. Tryon of Philadelphia was also contracted in 1848 to supply 5,000 rifles for the same unit price (Russell 2005). (Rock Island Auction)

lead to inaccuracy and wastage of ammunition: 'It is the opinion of the board that a larger proportion of fire from rapidly repeating guns would be thrown away, than from those that receive but one charge at a time' (Report 1837: 14).

Based on the success of the Hall percussion carbines, in 1839 the Ordnance Department asked Benjamin Moore, Master Armorer at the Harper's Ferry Armory, to design a new rifle with interchangeable parts and percussion ignition to replace all rifles then in service. Moore submitted several models, and in March 1841 the Ordnance Department selected the winner, later known as the M1841 'Harper's Ferry' or 'Mississippi' rifle. Moore prepared the Rifle Works for production and also made eight pattern rifles and gauges for use by private contractors. One pattern rifle was sent to Springfield Armory in Massachusetts to be used as a sample for the M1842 smoothbore musket.

The M1841's bore was .54in calibre, like that of the M1817 and all other 'common rifles', and it fired the same ball; but the charge of the cartridge could be reduced by 25 per cent because of the percussion ignition. The length of the bore was 33in (83.8cm) with six grooves and a twist-rate of one complete turn in 72in (182.9cm). The butt stock had a traditional patch box to accept the tools of the rifle and a spare nipple. This rifle was not equipped with a bayonet because – according to US military thinking of the time – the light infantryman did not need one.

Both armouries tooled up for production of the M1841, but work started slowly. The contribution of private contractors was once again essential. One of the first contracts was signed with the Whitneyville Armory in New Haven, Connecticut, led by Eli Whitney Blake, nephew of Eli Whitney, a key advocate of early mechanization. In 1842 an order was placed for 7,500 rifles at a cost of $13.00 each. Between 1842 and 1853, Whitney signed a further two contracts for 20,100 more rifles (Russell 2005).

In 1845, shortly before the beginning of the war with Mexico and following four years of debate in the United States concerning the general adaptation of the percussion system, the Ordnance Department decided to convert all existing flintlock firearms to percussion by drilling a hole on the right-side top of the barrel breech, threading it and screwing a nipple into it. The old touch hole was blocked off with a brass plug. The pan was sawn off the lock plate, and the hammer was replaced. Conversion of the Hall breech-loaders was easier: the pan was filed flat, the touch hole drilled and tapped for a nipple, and the hammer was replaced. The Ordnance Department accepted the modification method in 1845, but the solution was considered temporary.

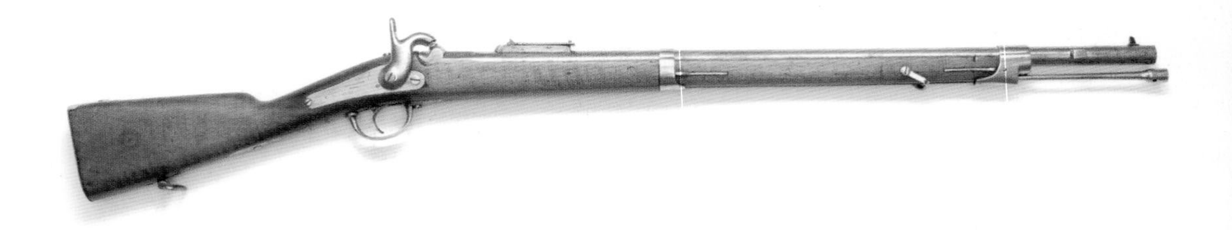

ABOVE & BELOW
This M1846 *carabine à tige*, produced in Belgium in the second half of the 1840s, has a Thouvenin-style pillar breech. (Author's collection)

French percussion rifles

One area of innovation after 1820 was the drive to replace the patched ball with an aerodynamically more efficient conical bullet. The first solution was provided by Capitaine Henri-Gustave Delvigne, who constructed a smaller-calibre chamber in the breech of his percussion rifle in 1828 (Götz 1978: 210). The undersized ball was loaded without a patch, and had to be hammered with the heavy ramrod to 'upset' it into the rifling for a tight fit. The concept was innovative but, as the form of the projectile varied from shot to shot depending on how hard the soldier hit the ramrod, the system was inaccurate. Lieutenant-colonel Charles Pontchara, also a Frenchman, tried to address the shortcomings of the Delvigne system. To avoid damaging the bullet Pontchara attached a wooden wad wrapped in a lubricated patch to the ball. This resulted in better accuracy, but the cartridge was overcomplicated and fragile. Even if the system was not perfect, the French Army adopted the 17.6mm Delvigne rifle and Pontchara cartridge in 1842, arming ten battalions of light troops with the new weapon.

Louis-Étienne de Thouvenin, an artillery major in the French Army, tried to solve the problem of symmetric expansion by adding a central pillar in the chamber of the rifle. The undersized projectile was easy to load, but still had to be hammered onto the pillar to expand it into the rifling. His percussion rifle – the *carabine à tige* ('stem rifle' or 'pillar breech rifle') – provided a much better solution than any of the previous concepts. The system was originally designed for round balls in 1844, but Delvigne understood the benefits of the new concept and constructed a cylindro-conical bullet for the Thouvenin rifle. The new rifle was accepted for service by the French Army in 1846, and many other European armies quickly followed the example.

The Thouvenin rifle was a huge leap forward towards the modern muzzle-loading rifle, but still had its limits compared to the old smoothbore weapons. Its accuracy was clearly much better, but the

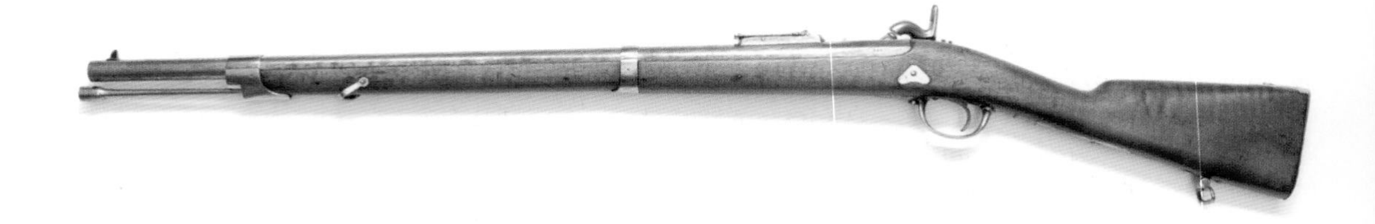

PERFECTING THE PATCH-ROUND BALL MILITARY RIFLE

The final effort to perfect the patch-round ball system is associated with Johannes Wild, a Swiss engineer and firearms constructor, who used a scientific approach to develop existing military rifles in the early 1840s. He set four principles as a benchmark. First, the bullet fired from a rifle had to achieve the high muzzle velocity of smoothbore musket ammunition to reach the same range. Second, the grooves had to be designed so that the high-speed bullet would not jump the rifling even with the highest powder charges. Third, the rifle had to be able to be loaded with cartridges without the use of a starter or mallet. Fourth, the rifle had to be capable of firing 100 shots without the barrel having to be cleaned of fouling.

Wild did not invent anything new, but used a scientific, mathematical approach to perfect the existing elements of rifle and cartridge. He reduced the twist-rate of the rifling to one turn in 102cm to allow the use of larger charges. He also changed the rifling profile. Instead of 6–8 grooves, he suggested the use of 16 grooves in a 16–17mm bore. His grooves were 1.9mm wide and 0.23mm deep, having a 3:2 width ratio between grooves and lands. He reduced the calibre of the ball, allowing a large windage of 0.7mm, but added an extremely thick (0.55mm) patch. This combination allowed easy loading without the necessity of a mallet or starter even in a fouled barrel. To control the fouling, he developed a previously unknown method: the patch of the cartridge was not lubricated, but was loaded dry. Wild developed a device with a special valve holding plain water that injected two cubic centimetres of water on top of the ball rammed into the breech. The water soaked the thick patch but did not harm the powder, so each shot actually wiped clean the fouling of the previous shot.

Wild's system proved superior to any previous methods. His rifle could be fired more than 100 times without cleaning and without losing accuracy; his projectiles reached 456m/sec, close to the performance of the smoothbore musket; and loading was easy even after many dozens of shots. During the trials, 30 shots were fired in 14 minutes, approaching the rate of fire of the smoothbore muskets. Even so, the system was adopted only by some of the smaller German states, including Württemberg, Baden and Hessen-Darmstadt. The future of the rifle lay in the development of the conical bullet (Götz 1978: 183–90).

hammering of the ball onto the pillar still reduced consistency in shooting, while the chamber was hard to clean. The solution to the problem was arrived at by a soldier serving at the French school of musketry at Vincennes, Capitaine Claude-Étienne Minié, whose system did not need the pillar in the breech as he modified the form of the bullet by hollowing the base. The skirt of the bullet could be easily expanded by the gas pressure. For a symmetric expansion he also added an iron cup, the *culot*, in the hollow base of the bullet, which was forced forward when the rifle was fired. Minié's system was perfected by Capitaine François Tamasier, who invented progressive-depth rifling. The original reason for having deeper rifling grooves at the breech and shallower grooves close to the muzzle was very simple: Tamasier intended to rifle the old musket barrels to save cost and time for the French Army. These barrels were very thin at the muzzle, however, so using traditional deep grooves was out of the question as it would have weakened the material to a critical degree. Although it was a simple makeshift measure, Tamasier's innovation, implemented from 1849, turned out to be a useful modification: the deeper grooves at the breech absorbed the residues of the powder better, while the shallow grooves at the muzzle aided a more gas-tight fit of the bullet.

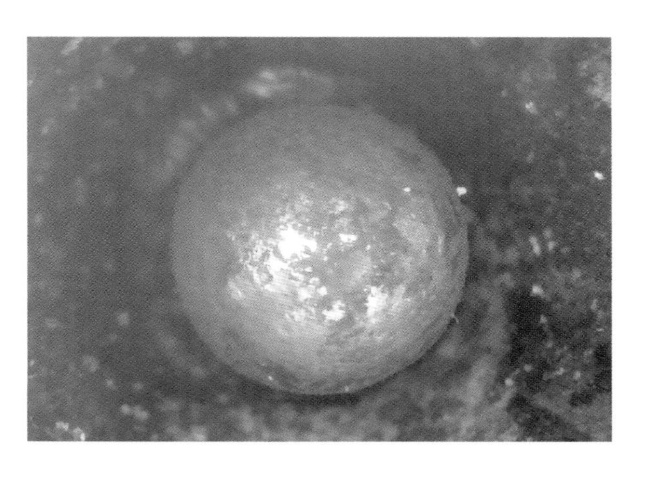

Taken with a digital borescope, this image shows the point of the pillar in the breech of the M1846 rifle. (Author's collection)

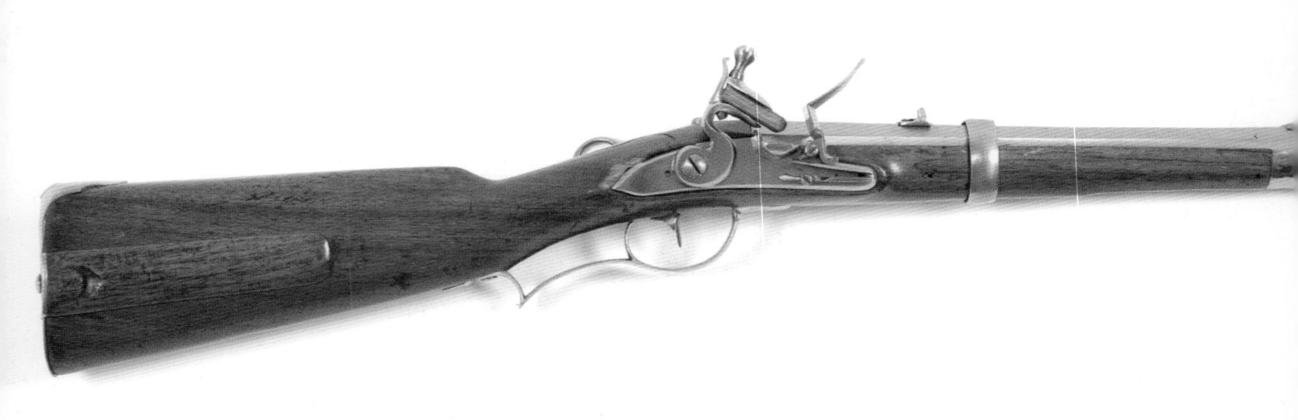

USE

The military art of the rifleman

ABOVE
The development of rifled weapons for the cavalry, represented by this Habsburg Army M1798 *Kavallerie-Stutzen* rifle in 5/4 Loth (15.6mm) calibre, proceeded alongside that of the infantry rifle and 'small war' tactics. Note that this example's rear and front sights are both in a turned position, probably as a result of restoration. (Museum of Military History, Budapest)

In the period 1740–1850 all of the major military powers deployed an assortment of regular units – infantry, cavalry and artillery – but by the end of the 18th century it became clear that regular light-infantry troops were necessary to support the operations of the line infantry. In the view of the Archduke Charles, a leading Habsburg military commander of the Napoleonic era, the line infantry was the only arm that could win battles; the light infantry played an essential role, however, in protecting and screening friendly line infantry from hostile activity by adopting open-order formations and harassing the enemy (Hapsburg 2009).

Owing to their irregular nature, before 1789 light troops were mainly used outside the battlefield. They (including those armed with rifles) had specialized tasks when an army was on the move or in camp – reconnaissance, screening the vulnerable marching columns and guarding encampments – which the line infantry were not able to fulfil effectively. Light troops also played an important part in fighting the 'small war'. A

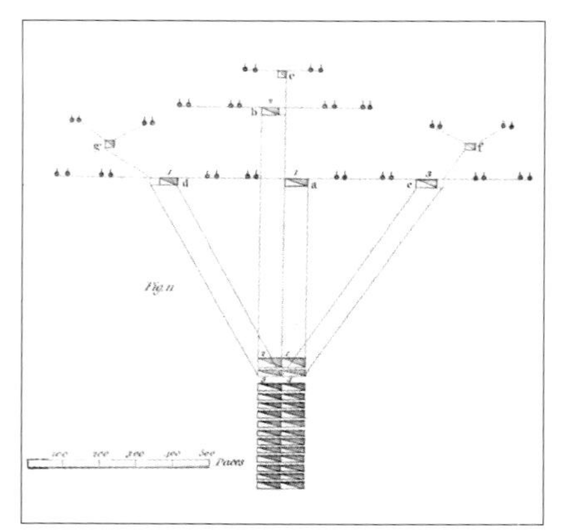

RIGHT
This illustration from a British manual shows 'the Formation and Disposition of a Company of Riflemen or Light Infantry which is to form an advanced Guard … the commanding officer marches with his half platoons to *a*, detaches 2 to *b*, 3 to *c*, & 4 to *d*; from *b* a non-commissioned officer is sent forward to *e*. From *c* to *f* & from *d* to *g* are the skirmishers' (War Office 1803). (Author's collection)

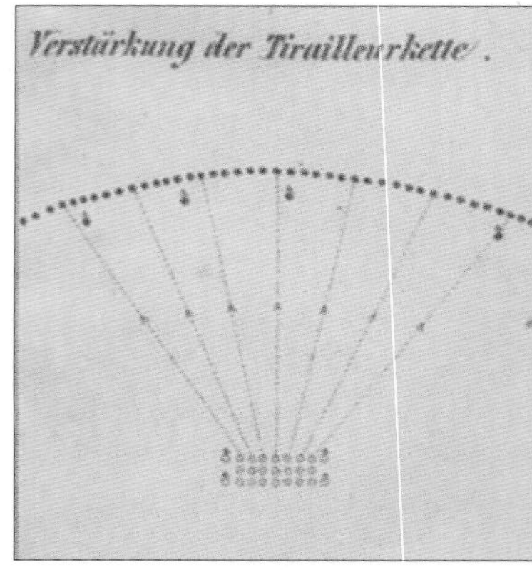

FAR RIGHT
'Reinforcing the tirailleur chain': this illustration was published in Anon 1851. The infantry tactics based on the close- and open-order combat formations of deployed lines, columns, squares and skirmish lines, developed during the French Revolutionary Wars and Napoleonic Wars, did not change in technical terms until the end of the muzzle-loading era. (Author's collection)

DISCIPLINE AND DRILL

In the period 1740–1850, there were two approaches to raising light troops. The first method was to select well-trained soldiers from line-infantry units and expand their development to become better-trained, more versatile light infantrymen who could use their initiative. The second method was the approach espoused by the nascent United States and Republican France, in which civilians were enrolled into the Army; as they were neither adequately trained nor had the necessary discipline, they were used mainly for light-infantry duties on and beyond the battlefield. The first solution resulted in a more effective soldier, therefore all armies sought to raise light troops based on former regular troops.

Those drill books intended specifically for the light-infantry soldier incorporated an inflexible drill in the training, for two reasons. The first was to instil military discipline and a sense of belonging, because the ordinary soldiers of most armies of the period were poorly educated and seldom felt any loyalty to the regime or understood the causes of the war. The use of the marching column and close-order combat formations such as the line and the square proved to be effective in preventing desertion. If a soldier was to leave these formations, or was detached from the main body of the army to join an advanced or rear guard, or was sent on a reconnaissance mission, or picket duty, or formed the loose chain on the battlefield, he had to have particularly strong discipline to ensure that he returned to his unit.

The second reason is closely connected to the conduct of revolutionary armies. Enrolling free, highly motivated civilian volunteers in the Army harnessed a formidable *élan*, but immediately generated a pressing problem: lack of expertise in the conduct of close-order warfare. The Patriot militias of the American Revolutionary War and the troops of the early French Republican armies shared a common difficulty: both included large numbers of civilian soldiers who understood the causes of the war and were willing to fight for their nation, but lacked the intensive close-order training required to face properly drilled line infantry, and so could be used only in open-order combat formations and for other classic light-infantry roles.

combined-arms detachment fighting the small war had to be able to operate in exactly the same way as did a large army, but on a smaller scale, utilizing a quicker, more flexible, proactive approach. Its tasks were various, excluding open clashes with the enemy main force, and included attacking enemy communications and logistics networks and operating in weakly garrisoned areas in order to compel the enemy to respond by dispersing his forces (Hapsburg 2009).

Light troops participated in 18th-century battles, but their role was limited as most were irregular troops. Moreover, deploying a battalion in open order during the 18th-century wars always carried the threat of desertion. The French Revolution (1789–99) created the circumstances to

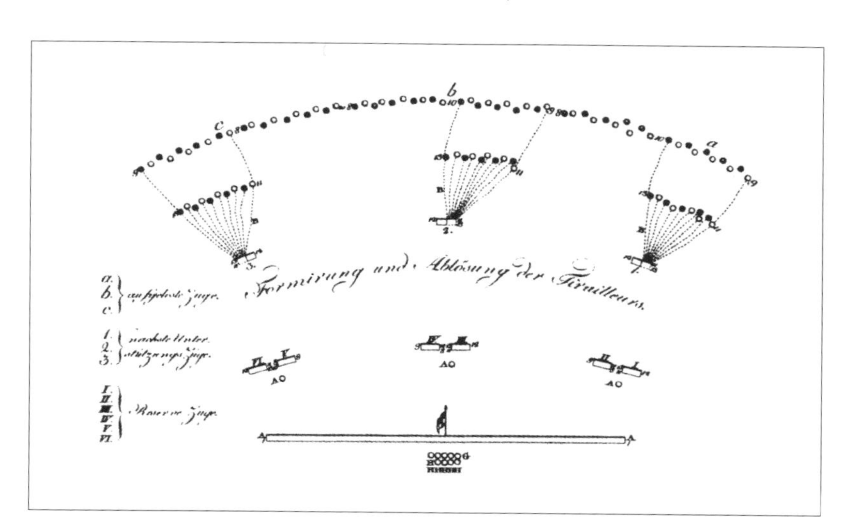

Taken from Auszug 1807, this illustration depicts the most advanced version of pulling the skirmish line or chain in front of the close-order combat formation; it depicts the chain itself, the support troops and the reserves. The skirmish line could attract enemy small-arms and artillery fire, thereby protecting friendly close-order troops nearby, while offering much less of a target than a close-order formation. Its marksmen could target enemy officers, NCOs and musicians, thereby spreading confusion, or could pin down enemy artillery crews. The skirmishers could fight on rough terrain and occupy strategic points of the battlefield quickly. They could hold the enemy at a distance to give time for friendly line infantry to deploy, and could also prevent friendly close-order troops from becoming involved in a firefight too early and against their will, or conceal the intentions and manoeuvres of the commander. The skirmishers could also pursue and harass the retreating enemy (Clausewitz 2015: 28). (Author's collection)

develop a more effective form of open-order combat. Although they were poorly trained, the soldiers of the new French Republic were motivated by patriotism and were therefore less likely to desert. Such troops could only be used in open-order combat formations or '*grandes bandes*', operating in front of the line-infantry formations. The skirmish line and close-order formations such as the attack column suited the French Army well as such tactics did not require much training and were more dynamic than the deployed line. Turning these soldiers into well-trained *tirailleurs* ('shooters') was achieved through practical battlefield experience rather than the application of theory. The French revolutionary army synthesized the different battlefield formations of the late 18th century and harmonized the fighting activities of the attack column, the deployed line and the skirmish line. Even among France's opponents, the skirmish line composed of pairs of light-infantry soldiers backed by support troops and reserves in close-order formation soon became the standard way to engage the enemy. The skirmishers usually did not exceed 4–5 per cent of the total strength, however; the fight was still decided by the close-order troops (Hapsburg 2009: 843–855).

The rifle-armed troops of Prussia, Austria, the United States, Britain and latterly France all developed their own doctrine during this period, as each sought to respond to the challenges of the battlefield as weapons technology and the nature of warfare evolved.

Taken from Gumtau 1835, this soldier of the Prussian Army's Garde-Jäger-Bataillon in 1834 carries the M1810 *Neue Corpsbüchse*. After 1821, the Garde-Jäger-Bataillon and the Garde-Schützen-Bataillon remained in service, but the other units were divided between the army corps as four *Jäger* and four *Schützen Abtheilungen* (divisions), each consisting of two companies (Gumtau 1835: II.4). (Author's collection)

PRUSSIAN RIFLE-ARMED TROOPS

In the mid-18th century the Prussian Army's cavalry arm included a regular light element, the hussars. By contrast, the light element of the infantry arm was made up of ad hoc units raised solely for the period of a conflict. Recruited and organized by military entrepreneurs who entered into contract with the Crown, these units were termed *Frei* ('free') battalions or regiments, as they were independent and formed separate administrative and tactical commands. Manned mostly by non-Prussians and even impressed prisoners, these battalions were armed with the standard infantry musket (Paret 1966: 31–32).

Conversely, Prussian Army *Jäger* were selected from foresters and hunters until the middle of the 19th century. These soldiers were close to the government administration, and they naturally fulfilled law-enforcement duties when countering the activities of poachers. They proved to be more loyal to the Prussian Army than did the regular line infantry, meaning that desertion was not a major issue, and therefore the *Jäger* could be better utilized in actions detached from the main force. The original unit was soon disbanded, but in 1744 the same approach was followed when organizing two 300-man volunteer companies, the Feldjäger-Korps zu Fuß. In 1773 the Feldjäger-Korps was reorganized into a five-company Feldjäger-Bataillon.

In 1784 the *Frei-Korps* units were reorganized into three ten-company light-infantry regiments, termed *Frei-Regimenter* and armed with smoothbore muskets, while the Feldjäger-Bataillon was reorganized into the Feldjäger-Regiment. Although Frederick II thereby increased the number of light troops, only 700 of them served with rifles, a further 600 soldiers receiving smoothbore muskets with bayonets. These smoothbore firearms were replaced with rifles in 1787 (Gumtau 1831: I.76–77), the same year in which the *Frei-Regimenter* were finally merged into the standing army (Paret 1966: 40). The soldiers of the new *Füsilier* battalions were armed with a shorter (134cm), better-made smoothbore musket, the M1787 *Füsiliergewehr*. Only ten soldiers of each *Füsilier* company – the *Schützen* – were armed with rifles (this number rose to 22 per company in 1798). Line-infantry regiments were also assigned ten *Schützen* per company in 1787.

When Prussia entered the French Revolutionary Wars, its riflemen were divided between *Jäger*, *Füsilier* and line-infantry units. The catastrophic defeats of 1806–07 and the humiliating Treaties of Tilsit of 1807 led directly to the full-scale reforms of the Prussian Army that followed. In the event, the *Füsiliere* became part of the line infantry, each battalion being assigned to a line-infantry regiment and adopting the blue line-infantry uniform. In 1808–09 the Prussian Army's rifle-armed troops were reorganized into three battalions: the Garde-Jäger-Bataillon, the Silesian Schützen-Bataillon and the East Prussian Jäger-Bataillon. By 1815 another three battalions had been added to the Prussian Army's order of battle: the Garde-Schützen-Bataillon, a third Jäger-Bataillon and the Rhenish Schützen-Bataillon (Gumtau 1831: I.231). In addition to these units, several thousand volunteer riflemen, each being required to arm and equip himself at his own expense, were attached to the line-infantry regiments.

Prussian rifle training

Although the Prussian Army was one of the first to realize the benefits of regular rifle troops, it did not pay enough attention to proper training. According to Carl Friedrich Gumtau, a senior officer and *Jäger* veteran, before 1806 only nine shots per year were permitted for target practice, and lead from used projectiles had to be collected and recast into new balls (Gumtau 1834: I.59). Certainly, the Prussian Army relied on the skills the riflemen acquired before enlisting. Once Scharnhorst's Commission started its work in 1807, however, marksmanship finally became one of the most important elements of training. Each year, every Prussian infantryman received powder and lead for 60 shots during the summer exercises, and during winter each group of 202 soldiers received an allowance of ½ Zentner (25.7kg) of gunpowder, enough for an additional 20 shots each. For the spring and autumn manoeuvres held after 1815, an additional 1 Pfund (467.7g, 7,218 grains) of powder was issued to each soldier of the Garde-Jäger-Bataillon and Garde-Schützen-Bataillon and ¾ Pfund (351g, 5,417 grains) of powder for each soldier of the *Jäger* and *Schützen-Abtheilungen*. This permitted another 60–80 shots per year, resulting in 140–160 shots per soldier per year (Gumtau 1835: II.50–51). Even so, the yearly lead allowance was still limited to 60 bullets, so recycling the spent rifle balls remained a necessity.

Using a rifle's sights required skill in judging distances and an intelligent mind to select the right combination for each distance. This diagram, taken from Gumtau 1835, shows the three methods for placing the bead into the rear sight. Prussian soldiers were taught three sight settings: *fein Corn*, when the top of the front sight was just visible at the bottom of the notch of the rear sight; *gestrichen Corn*, when the top of the front sight was level with the top of the rear sight; and *voll Corn*, when the front sight was elevated above the top of the rear sight (Gumtau 1835: II.136). (Author's collection)

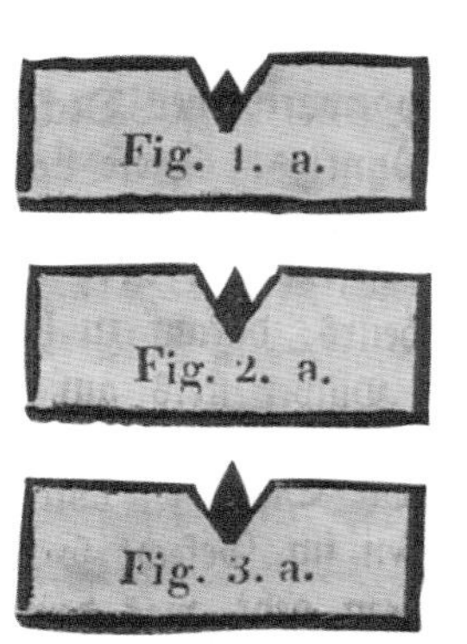

The training of a rifleman started with instruction in the basic principles of rifle shooting, and learning the correct management of the rifle. The second step was to learn the proper hold: the rifle was placed on the shoulder and the right elbow raised to the same height as the shoulder. The cheek was placed on the rifle gently without tilting the rifle. The left hand supported the rifle close to the centre of gravity. This comfortable, balanced hold could be tested by closing both eyes, then placing the cheek on the buttstock. If, when opening the eyes, the hold was good, the sights offered a good sight picture. New recruits without shooting experience became accustomed to this by dry-firing the rifle with a primed pan. The novice had to keep his eyes open regardless of the flash in the pan.

The rifle was inspected just before target shooting. The first charge to be fired was a small amount of powder to test whether the vent hole was clean or not. For practice sessions the soldiers avoided the use of iron ramrods, as they could damage the rifling during intensive use. A small piece of tow was placed in the pan to prevent the powder of the main charge spilling out through the touch hole. This was necessary even if the rifle had a French-style conical touch hole, not the traditional self-priming type of previous Prussian firearms. Then the main charge was carefully poured into the bore, with the barrel held vertical. After pouring in the charge the shooter tapped the barrel a few times to help the powder settle in the breech. Then the lubricated patch was placed on the muzzle. The bullet was placed on the patch and pushed into the bore. The rifle was lowered between the legs with the barrel towards the body and the ball was pushed down firmly to the powder, but without crushing the charge. The proper position was checked by dropping the ramrod on the charge; if it bounced back, the procedure was correct. The riflemen marked the ramrod now to be able to check the proper seating depth later (Gumtau

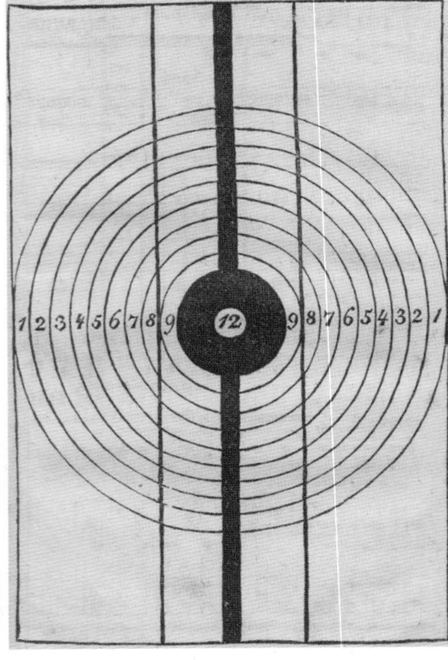

RIGHT
This rifleman of the Prussian Army's Garde-Jäger-Bataillon, pictured *c.*1815, is still equipped with an *Alte Corpsbüchse* lacking the means to attach a bayonet. Not all rifles were the same, so finding the right load for each rifle was also a part of the training. According to Gumtau this charge could be as large as the soldier was able to tolerate in terms of recoil, but based on his experiences ⅓ Loth (4.9g) of powder was usually enough. Shots taken at different distances had to use the same charge, so even if the powder measure was adjustable, increasing the charge for longer ranges was not allowed, as doing so was likely to make the rifle less accurate. (Author's collection)

FAR RIGHT
Taken from Gumtau 1835, this circular target entered service with the Prussian Army at the request of the Scharnhorst Commission, to aid a better understanding of the principles of accurate shooting. The board was 4ft (1.26m) wide and 6ft (1.9m) high with a circular target painted on it; 12 circles were drawn at 2in (5.23cm) intervals, resulting in a 125cm-diameter bull's-eye with rings scored from 1 to 12. The shooter's vertical hold was aided by a horizontal black line, 2in (5.23cm) wide, across the target (Gumtau 1835: II.127–31). Target practice was initiated at 100 paces (75m). Five shots were fired and an average score of at least 8 was to be achieved before proceeding to 150 paces (113m). At this range, the five shots had to achieve an average score of at least 7; at 200 paces (150m), 6; at 250 paces (188m), 4; and at 300 paces (225m), an average score of 2 per shot (Gumtau 1835: II.127–33). Soldiers were grouped into four classes based on their shooting skill, and all results were recorded in the company shooting books to track improvement. (Author's collection)

PRUSSIAN RIFLE AMMUNITION

Each rifleman armed with the M1810 *Neue Corpsbüchse* carried balls and powder for 60 shots, divided into three types: 20 standard-calibre *Pflasterkugel* (patched ball) charges with patches in a leather pouch and loose powder in the powder horn; 20 paper cartridges holding a smaller-calibre *Patronenkugel* (cartridge ball) and the powder charge, placed in the cartridge box; and another 20 paper cartridges in the cartouche with *Pflasterkugel* loaded without the patch. Supply for a further 30 shots was transported on the ammunition carriage.

Lead and powder were supplied from the artillery depots (Gumtau 1835: II.46). The bullets of the cartridges had to be loose fitting, so they could be pushed into the muzzle with the thumb. Cartridges were bundled together in packets of ten, with white wrappers for the *Pflasterkugel* cartridges and blue for the *Patronenkugel* cartridges (Götz 1978: 164). The powder allowance for one shot was 5/12 Loth (6g) of *Pirsch Pulver* (hunting powder), corned powder of a finer grade than the standard musket powder: 1½ Loth (21.9g) of lead was calculated for one shot. The powder was measured by the soldier himself using an adjustable powder measure. The scale of this device was divided into 12 grades. It measured the proper 5/12 Loth (6g) charge set to Grade 8. If the soldier was making cartridges, he set the scale to Grade 9 to compensate for the loss of powder when biting off the end of the paper case. The *Pflaster* (patch) of the M1810 rifle's cartridge was circular; it was made of good-quality cloth and was dipped into a mix of tallow and pork fat. Bullets were cast with the bullet mould supplied with each rifle. Each company had a supplementary mould as well, casting 12 slightly different-calibre balls so matching projectiles could be found for each rifle in the company (Gumtau 1835: II.52).

Each soldier was supposed to roll his own cartridges, using a wooden dowel which was designed to enter the bore easily when the paper was wrapped around it. The length was 6–8 Zoll (15.7–20.9cm), with one end hollowed for the bullet. The trapezoid of the cartridge paper measured 4⅓ Zoll (11.3cm) in length, 4½ Zoll (11.8cm) high where the bullet was placed and 2½ Zoll (6.5cm) at the tail of the cartridge. The patch's diameter was 2–2½ Zoll (5.2–6.5cm), so it could wrap the ball completely and be 'choked' with a thread. The paper case was then rolled on the wood and the patched ball inserted into the tube where the hollowed end of the dowel was. The paper was 'choked' on the tail of the patched ball, then with the bullet towards the table it was firmly pushed deeper into the tube to compress the tail. The ball was now immersed in the molten lubricating mixture, and the powder case filled with the proper charge. The tail of the case was then folded to close the cartridge. The final step was to cut a cross on the patch with ⅓ width to facilitate the patch separating from the projectile after leaving the muzzle (IKBPJCB: 19–21).

1835: II.136). When the rifle was loaded with cartridges the soldier grasped and flattened the paper case close to the top of the powder column with his thumb and index finger and bit off the folded tail. The opened end of the cartridge was now inserted into the muzzle to let the powder pour into the bore. The bullet with the paper sleeve was now pushed into the muzzle with the thumb and rammed down with the ramrod. In both cases the final stage of the process was to prime the pan from the powder flask (IKBPJCB: 24).

After each use the rifle was to be cleaned immediately. The barrel was removed from the stock and the bore was filled several times with clean warm water to soften the fouling. Then the barrel was half-filled again and the breech placed in water. A wooden cleaning rod wrapped with tow at the end was used to clean the residue of gunpowder until the tow remained clean. The water was then poured from the bore, and the barrel was then dried with tow. The dried barrel was oiled and replaced in the stock. The lock was to be cleaned and oiled as well, but without taking it apart completely. The stock was also cleaned, and frequently rubbed with a rag saturated with linseed oil to protect it from moisture (IKBPJCB: 15–17).

Prussian rifle tactics

Owing to their irregular nature, the *Frei-Korps* were not suitable for fighting in close-combat formations, while the *Jäger* were originally intended to fight mainly in open order. According to Frederick II, a *Frei-Bataillon* – considered to be a second-line unit – could participate in the first wave of the attack; it was to attract the enemy's fire to allow the line regiments behind it to prepare for the decisive charge (Paret 1966: 35). The *Frei-Korps* recruits were not trained for skirmishing or marksmanship, and did not have the proper firearm for such tasks. Although the *Jäger* armament was more suitable for accurate shooting, on a tactical level they shared the same principles as the *Frei-Korps*.

The evolution of the *Feldjäger-Korps* is strongly connected to Philipp Ludwig Siegmund Bouton des Granges. Born in Switzerland in 1731, he started his military career in Dutch service, but in 1757 enlisted in a Prussian *Frei-Regiment*. He became the commander of a company of the *Feldjäger-Korps zu Fuß* in 1759 but was subsequently captured near Spandau when the complete *Jäger* force was ambushed and destroyed by Cossacks as it crossed a field, showing the vulnerability of rifle-armed troops to sudden fast attacks. At this time the Prussian Army's *Jäger* did not carry a bayonet, which could have been helpful for the defence. Des Granges returned to his unit in 1762 and became its commander in 1773. Following his unit's participation in the War of the Bavarian Succession (1778–79), Des Granges worked intensively on the further development of the *Jäger* troops.

In December 1783 the first light-infantry instructions were published, titled *Instruction für die Frei-Regimenter oder leichten Infanterie-Regimenter*. The *Schützen* received their own instructions in 1789, titled *Instruction für sämtliche Infanterie-Regimenter und Fusilier-Bataillone. Exercieren der Schützen betreffend*. According to the regulations, the sharpshooter had to be receptive to new ideas and have the potential to become a non-commissioned officer; only those who had served as sharpshooters could be so promoted (RKPI 1789: 186). The new light-infantry regulations allowed fighting in open order, and the *Füsilier-Bataillone* were formed in two ranks, not three as in the case of the line-infantry regiments. The two-rank formation facilitated the loading process and meant it was easier to keep the unit in line in broken terrain; it could be deployed into extended order more quickly and offered a less vulnerable target (Paret 1966: 58).

RIGHT
These soldiers of the Garde-Jäger-Bataillon, *c.*1835, are equipped with percussion rifles. Percussion caps were first manufactured by the Sellier & Bellot factory in Schönebeck an der Elbe (Gumtau 1835: II.47). Each soldier was to receive 150 caps in wartime, which he kept in a separate compartment of the belly cartridge pouch (Gumtau 1835: II.50). This compartment was closed by a leather flap padded with fur to seal the pocket from moisture. (Author's collection)

FAR RIGHT
Published in Gumtau 1838, this fascinating image shows two soldiers of the Garde-Schützen-Bataillon armed with the M1810 *Neue Corpsbüchse*. In this picture, Gumtau shared his view about the future equipment and uniform of the Prussian Army's light infantry, prefiguring important developments fulfilled when the M1841 *Zündnadelgewehr* was introduced. He suggested exchanging the frock for a single-breasted tunic, the adoption of a looser style of legwear to ease movement, and the replacement of the shako with a leather helmet. He also advocated important changes to the leather accoutrements. The cartridge pouch was to be worn in front of the belly, following the curve of the body, with the compartment system redesigned to aid access to the components of the load. The backpack should be as small and light as possible; the backpack straps were connected to the waist belt, which held the cartridge pouch and the bayonet scabbard. This arrangement offered a more balanced distribution of weight, while eliminating the tight chest strap facilitated the soldier's breathing and increased his endurance. Gumtau also suggested replacing the brass powder horn with a waterproofed leather bottle. (Gumtau 1838: III.372–89). (Author's collection)

PRUSSIAN RIFLE ACCESSORIES

Accessories for the M1787 *Schützengewehr* included a cylindrical steel ramrod, a leather cover for the frizzen, a leather lock cover, a ball puller, a scraper and a leather *Brandriemen* (fire strap). This last item was intended to protect the left hand from the heat of the barrel during rapid firing, indicating that even if the weapon's bore was rifled, the Prussian Army still paid more attention to a high rate of fire than to accuracy.

When the Prussian Army's first new rifle units were established in 1808–09, they lacked good-quality rifles. The old rifles were in poor condition and mostly unserviceable. As a makeshift measure, hunters and foresters were asked to hand their good-quality rifles to the *Jäger*, to be returned after the new rifles arrived. In February 1813, when Prussia joined the Sixth Coalition against Napoleon, still only a small proportion of the rifle-armed troops was armed with the new rifles; many still had the old models, or had to serve with civilian firearms. Regardless of the type, every rifle was equipped with a bayonet rail to accept the 68.8cm-long brass-handled *Hirschfänger* ('deer catcher') – a straight double- or single-bladed hunting sword used in Central-Eastern Europe as the self-defence arm for big-game hunting – weighing 1.08kg (Gumtau 1834: I.229). The blade was 3.26cm wide.

The rifleman was equipped with the same cartridge box as the line infantryman, but he also carried a smaller leather cartridge pouch for bullets, patches, cartridges and a small metal flask of oil. This pouch was made from soft black leather and was worn in front of the belly. Its internal space was divided into smaller compartments. The powder horn used by the riflemen was made from strong brass sheet. It had a convex form on both sides and was large enough to hold 20 Loth (292g) of gunpowder. The spout of the flask was sized to be used as a volumetric powder measure for the charge. The flask hung on a narrow leather belt over the shoulder in combat, but during marches and parades it was kept in the pocket (Gumtau 1835: II.36–37). Each rifle was equipped with a leather sling and a leather lock cover to protect it from the rain. The cover attached to the rifle with straps. The part covering the rear sight was padded with fur for additional protection.

The brass-handled *Hirschfänger* used with the M1810 *Neue Corpsbüchse*, the Prussian Army's first military rifle to be equipped with a bayonet. (International Military Antiques ima-usa.com)

Even though Prussia had been the first country to organize *Jäger* units, tactical innovation in the Prussian Army slowed down by the end of the 18th century. Before 1806 the Prussian light troops' performance was inferior to that of Austrian, British and US rifle-armed troops – and senior figures in the Prussian Army realized this. According to contemporary Prussian sources, the *Jäger* and *Füsiliere* were trained much like the line infantry, the *tirailleur* ('shooter') system was little known and the riflemen lacked training in extended order and proper target practice (Paret 1966: 60).

In 1812 each of the Prussian Army's combat firearms received new regulations edited by Hans David Ludwig von Yorck, Graf von Wartenburg, implementing French-style tactics and finally ending the practice of using the light troops as the first wave of the attack. The light-infantry soldier was primarily to conduct the firefight, and to refrain from engaging in close combat; the *tirailleur* was to protect himself from being hit, using the cover offered by the terrain (Paret 1966: 159). Adaptation to the tactical situation and harnessing the benefits of the terrain required a more intelligent approach to skirmishing, combining fire, movement and intelligent use of cover (Paret 1966: 167). The new regulations also implemented the theory of the dual-purpose infantryman who was able to act as *tirailleur* or line infantryman as needed (Paret 1966: 163).

RIGHT
This coloured lithograph by Franz Gerasch depicts a Habsburg Army *Grenz Scharfschütze* at the end of the 18th century. The *Grenz Scharfschütze* carried his rifle in a *Schützensack* (large leather bag) to protect it from the elements. A small leather pouch was sewn to the side of the bag to store 12 brass loading tubes (see page 45). The ramrod was attached to the strap of the *Schützensack*. (Author's collection)

FAR RIGHT
In this illustration by the military painter Rudolf Otto von Ottenfeld (1856–1913), published in 1895, a *Grenz Scharfschütze* (left) is accompanied by a musket-armed comrade in the Habsburg Army uniforms worn before 1798. The *Döppelstutzen* rifle was not equipped with a bayonet; instead, the sharpshooter received an infantry sword and a 2.53m-long lance for close combat. The lance had an iron hook and could also be used as a support for the rifle. (Author's collection)

HABSBURG RIFLE-ARMED TROOPS

The Habsburg *Grenzer* troops formed the largest continuously serving body of light troops among all the leading military powers of the 18th century. Families were selected for military duty, which was passed from father to son; they served in exchange for partial exemption from taxation. To increase the number of regularized light troops, the Archduke Charles transformed the *Frei-Korps* into 14 light-infantry battalions in 1798. Later, parts of these were merged into the Tyrolean *Jäger* regiment raised in 1801. Two-thirds of the men – *Carabiner-Jäger* – serving in these battalions were armed with smoothbore infantry muskets and one-third – *Stutzen-Jäger* – with the new M1795 *Jägerstutzen* rifle (Sitting 1908: 21). After the Treaty of Schönbrunn (14 October 1809), the Archduke Charles continued the reforms and established the system of *Feldjäger-Bataillone*, 12 of which were organized up until 1813. To replace the *Frei-Korps* he established the *Landwehr* volunteer system and increased the proportion of light troops.

Habsburg rifle training

Although the Habsburg Army tended to recruit light troops from men already accustomed to using the rifle, not all soldiers shared the same skills in marksmanship. Target practice was used to decide whether a recruit received a rifle or smoothbore musket. The regular distances for individual target practices were 150 paces (112m) and 200 paces (150m) with circular targets, while volleys were practised on wooden planks that were 6ft (1.9m) high and many metres wide (Paumgartten 1802: 72–73). From 1807 each *Stutzen-Jäger* received a yearly allowance of 20 *scharf* ('sharp', i.e. bulleted) cartridges, raised to 35 in 1819 (Götz 1978: 176). Target practice and range judgement were particularly important as the trajectory of the M1807 *Jägerstutzen* rifle was markedly curved. The soldier had to learn the use of fine, full and elevated sights, with both the block and the leaf sight. The longest range at which the rifle could operate was 350 paces (262.5m), using elevated front sight and leaf rear sight.

The brass loading tube for the *Doppelstutzen* rifle (reproductions are shown here) had separate compartments for the patched ball and the powder, closed with a felt wad. (Author's collection)

HABSBURG RIFLE AMMUNITION

Loading the Habsburg Army rifles took longer than loading a substantially undersized round ball into a smoothbore musket. The Habsburg Army developed several ways to speed up the process, as it did not issue *Rollkugel* cartridges to its riflemen. The most common way to load the rifle for maximum accuracy was to charge the bore with loose powder measured with a volumetric powder measure from the powder horn. The horn itself did not have an adjustable valve, just a plug that held a pin for cleaning the touch hole. During the late 18th century the rifleman used 1 Quintel (4.375g) of *Scheibenpulver* (fine rifle powder) and a 14.16mm-diameter round lead ball wrapped in a triangular lubricated linen or leather patch (Unterberger 1807: 47). To facilitate the loading process the powder charge could also be wrapped in a paper cartridge, but the ball and patch were still carried separately. The calibre of the Habsburg Army rifles was reduced to 13.9mm in 1807, but the cartridge specifications remained the same, rendering the fit of the projectile in the bore tighter and increasing the force necessary to start the ball in the muzzle.

Another method, probably the most expensive of all, was to issue brass loading cartridges. The *Grenz Scharfschütze* received 12 tubes, each with a separate compartment on either side – one holding the powder charge closed with a felt wad and the other holding the round ball wrapped in the triangular patch. The corners of the patch protruded from the compartment, so the ball was easy to grasp and pull. These tubes could be lost easily in the heat of battle, and were withdrawn from service by the end of the 18th century (Gabriel 1990: 39).

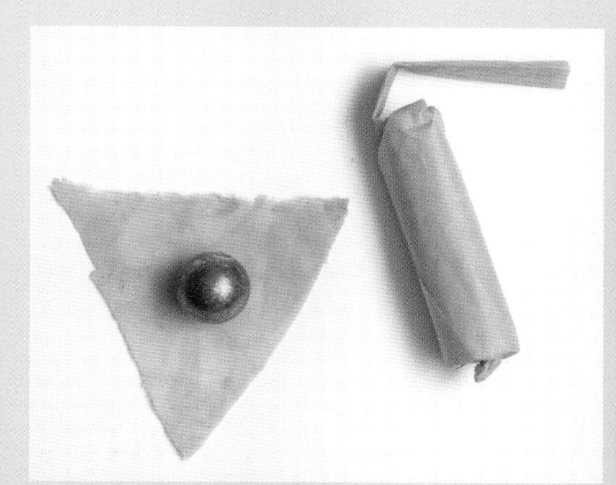

A reproduction of the *Jägerstutzen* rifle cartridge, with powder charge enveloped separately in a paper tube. The ball and triangular, lubricated patch were carried separately (Hoyem 2005: 13). (Author's collection)

A reproduction cartridge for the M1842 *Kammerbüchse*, holding a patched round ball and the powder charge. (Author's collection)

RIGHT
This coloured lithograph by Franz Gerasch shows a Habsburg Army *Stutzen-Jäger*, 1809. (Author's collection)

FAR RIGHT
Taken from Anon 1851, this circular target integrated into a human body target is for *Jäger* shooting practice. (Author's collection)

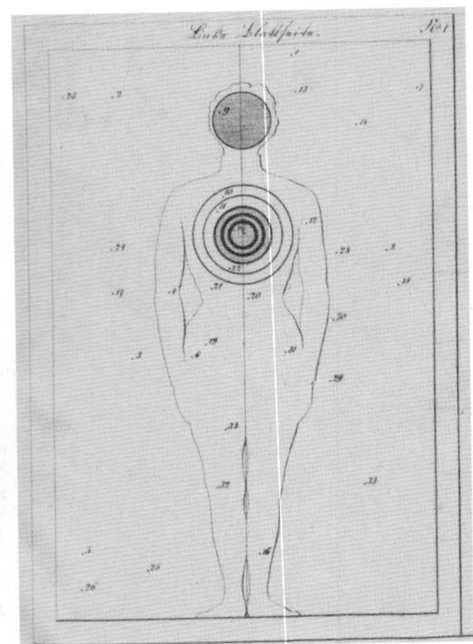

Unterberger advocated a more structured programme of marksmanship training for the rifle-armed *Jäger* (Unterberger 1807: 66–68). Target practice was initiated at 50 paces (37.5m) where the soldiers practised with half powder charges to learn the basics of musketry properly. The next stage was conducted at a range of 100 paces (75m), using a paper target that was 6ft (1.9m) high and 2ft (0.63m) wide and displayed a soldier. The recruit had to pass this stage without any misses. The next distance was 200 paces (150m), with the target being 6ft high and three paces wide (1.89×2.25m). The well-trained *Jäger* had more hits than misses at this distance. The third and last stage took place at a range of 300 paces (225m) using the same target with the same expectations.

Habsburg rifle tactics

The *Doppelstutzen* rifle was operated in 21 motions (Dolleczek 1970: 79):

1. *Macht euch fertig!* – Prepare.
2. *Ergreift die Patron!* – Grasp the cartridge.
3. *Zieht den Propfen aus den Patron!* – Pull the wad from the cartridge.
4. *Wischt damit die Batterie, Pfanne und Stein!* – Clean the frizzen, pan and flint with the wad.
5. *Gebt das Batterie-Futteral auf!* – Place the leather protector on the frizzen.
6. *Schwenkt zur Ladung!* – Make a half turn for loading.
7. *Schüttet das Pulver ein!* – Charge the powder into the bore.
8. *Stosset ein paarmal auf!* – Hit it a few times.
9. *Setz den Propfen darauf!* – Place the wad into the bore.
10. *Kehrt die Patron um!* – Turn the cartridge.
11. *Ziehet am Zipfel des drieckigen Pflaster!* – Pull the edges of the triangular patch.
12. *Nehmt die Kugel in die Hand!* – Grasp the ball.
13. *Steckt sie in den Lauf!* – Put it in the bore.

HABSBURG RIFLE ACCESSORIES

The accessories carried with the M1795 *Jägerstutzen* rifle included the powder measure, powder flask, leather lock cover, rifle sling, wooden mallet, leather pouch for patches and balls, cartridge box, ramrod and the bullet mould. To facilitate access to the components of the charge, many *Jäger* attached the patches to the hat cord. The cartridge box was larger than the line infantryman's cartridge box. It held the pre-rolled cartridges, the bullet mould, cleaning tools, a screwdriver, a little flask of oil and a few linen patches. The shoulder strap of the *Stutzen-Jäger*'s cartridge box had a two-ended leather strap attached to it, one end holding the powder measure and the other having a ring for the ramrod. Austrian *Jägerstutzen* rifles, cavalry carbines and holster pistols did not have a channel in the stock for the ramrod; instead the rammer was attached to the strap to facilitate loading. A leather strap on the cartridge box held the mallet for tapping the ball into the muzzle. On the other side a small leather pocket for spare flints was attached. The soldier wore the cartridge box on his right side, and the black leather scabbard for the sword-bayonet on his left.

The Habsburg Army's *Stutzen-Jäger* carried loose powder in the powder horn, but also carried pre-rolled paper cartridges filled with 4.3g of *Scheibenpulver* in the cartridge box. The 14.2mm lead round balls and the triangular patches were carried in a leather pouch (Dolleczek 1970: Tafel XVII). The cartridge allowance for each private was 60 and powder and ball for another 40, while NCOs received only 30 each. Each platoon had two lead ladles and two pairs of pliers for cutting the sprue of the bullets. Each company received a powder-testing device to verify the quality of the powder.

Austrian *Jäger* powder horn with probably the battalion number carved into it. (Author's collection)

Adjustable volumetric powder measure, retrieved from the patch box of an M1769 *Jägerstutzen* rifle. (Author's collection)

The *Hirschfänger* was synonymous with the German *Jäger*. The double-edged hunting sword was the last line of defence for the hunter and became the symbol of the hunter and the rifle troops in the 18th century. (Author's collection)

The sword-bayonet for the M1795/96 *Jägerstutzen* rifle was the first *Hirschfänger* to receive a socket mounting. In 1796, many of the older rifles' octagonal bores were reshaped for the final 11cm to accept the new bayonet. (Author's collection)

RIGHT
This illustration by Rudolf Ottenfeld, published in 1895, depicts a Habsburg Army *Stutzen-Jäger*, 1798–1805. The pike-grey uniform offered a better camouflage than the line infantryman's white clothing. The grass-green padding later sewn to the shoulders of the frock prevented the sling from slipping down when the rifle was slung muzzle-down on the left shoulder. The *Jäger* had to use both hands while moving quickly through rough terrain, so this feature was very helpful. From 1798 the *Jäger* wore the leather helmet, later changed to the *Korsehut* (Corsican hat). (Author's collection)

FAR RIGHT
This coloured lithograph by Franz Gerasch shows a Habsburg Army *Grenz Scharfschütze* armed with an M1842 *Kammerbüchse*. (Author's collection)

14. *Steck die Patronenhülse wieder ein und ziehet den Ladstock aus dem Ring!* – Put away the cartridge and pull the ramrod from the ring.
15. *Schlagt mit dem Kopf des Ladstockes die Kugel ganz hinein und stoßt sie mit dem Setzer nach!* – Hammer the ball into the bore completely with the head of the ramrod, then push it down with the starter.
16. *Macht euch fertig!* – Prepare.
17. *Nehmt des Batterie-Futteral ab und greift nach dem Pulverhorn!* – Remove the frizzen protector and grasp the powder horn.
18. *Schüttet das Pulver auf die Pfanne!* – Prime the pan.
19. *Blast ab!* – Blow (the excess powder) off.
20. *Schließt die Pfanne und spannt vollends den Hahn!* – Close the pan and cock the hammer.
21. *Schlagt ruhig an!* – Calmly aim the rifle.

During the Napoleonic Wars, if a Habsburg Army *Feldjäger-Bataillon* was deployed in a three-rank line the *Stutzen-Jäger* formed the rear rank, while the musket-armed *Carabiner-Jäger* fought in the other two ranks. One attempt to harmonize their loading sequence was described by Sigimundus Paumgartten, who detailed an 11-stage process suitable for both types of soldier (Paumgartten 1802: 55–76). Paumgartten's method

Habsburg Army *Jäger* on skirmish duty, 1809 (opposite)

This plate shows a pair of *Feldjäger* in action against Napoleon's forces. One man, a *Carabiner-Jäger*, reloads his M1798 smoothbore musket while his colleague, a rifle-armed *Stutzen-Jäger*, kneels to achieve a better stance for accurate firing. The rifleman's *Jäger* cartridge box is equipped with a holder for the wooden mallet. The ramrod and powder measure are attached to the strap with leather strips. His powder horn hangs on grass-green cords with two large tassels. The *Carabiner-Jäger*'s shortened infantry musket is equipped with a socket bayonet, its frog holding the infantry sabre as well.

cancelled out the benefits of both types: it decreased the firing rate of smoothbore firearms and also eliminated the flexibility of the rifleman.

In 1806, new exercise manuals were adopted which incorporated French-style light-infantry tactics. Three ranks deep, the deployed line was considered to be the most capable infantry formation for both attack and defence. The tallest soldier stood in the front rank, the shortest in the second and the most experienced in the rear rank. The first two ranks fired the volleys while the soldiers of the third rank filled the places of wounded and killed soldiers, handed loaded firearms to the front and second ranks and refilled their cartridge boxes with ammunition. These third-rank soldiers could also be used for extending the flanks of the line to increase the number of firearms firing at the enemy (Auszug 1807: 37–39).

The most important task of the third rank, though, was to form the chain or skirmish line and to fight in *zerstreuten Fechtart* (scattered order). To form the chain in front of an infantry battalion, three half-platoons – 60–80 soldiers in all – were enough. The soldiers formed pairs. The distance between the close-order combat formation and the chain was 300 paces (225m). Behind the chain the support troops were placed in a two-deep deployed line formed from the other halves of the three platoons. The half-platoons' role was to protect the chain against sudden attacks and to fill the gaps if soldiers were wounded or killed. They could also relieve the half-platoons of the chain.

Another three platoons formed the reserve, deploying in a two-deep line formation 100 paces (75m) behind the support troops. When the chain retreated, it formed a two-deep scattered line. The soldiers withdrew in an organized manner, while maintaining their fire. While the first rank fired, the rear rank retreated ten paces (7.5m) and prepared to fire while the first rank retreated. When the distance from friendly troops closed to 50 paces (37.5m), both ranks turned and retreated to the sides of the friendly close-order formation in double step.

When the chain was composed solely of riflemen, groups of three rather than pairs were formed to compensate for the slow firing rate of the *Jägerstutzen* rifles. Each group of three was led by an experienced *Jäger* who stood in the rear, always keeping his rifle loaded for emergencies; the two other members of the group of three were responsible for accurate rifle fire, utilizing all cover offered by the terrain and firing their rifles from any positions (Sitting 1908: 21).

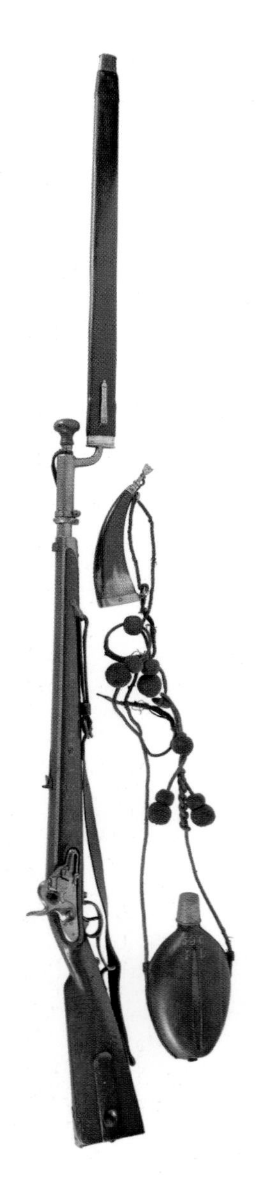

This M1842 *Jägerstutzen* rifle has the ramrod inserted into the barrel and is accompanied by a powder horn. (Dorotheum Vienna, auction catalogue, 12.11.2014)

In this illustration taken from Anon 1851, Habsburg Army *Jäger* form the *Klumpen* ('knot') to defend against an enemy cavalry charge after the friendly skirmish line has been broken. These *Klumpen* slowly retreated to the flanks of a unit of friendly troops. (Author's collection)

This flintlock rifle was made in Northampton County, Pennsylvania. (Rock Island Auction)

US RIFLE-ARMED TROOPS

Although the flintlock rifle is now considered to be a quintessential symbol of freedom, the smoothbore musket was by far the most numerous firearm in George Washington's Continental Army, and the role of the rifle and the impact of rifle-armed troops was greatly exaggerated subsequently. Even so, the first units raised for the Continental Army were armed with rifles. In 1775 the US Congress authorized the establishment of ten rifle companies to act as regular light infantry. These were supplemented by 13 more companies (Hess 2008: 21). Many of these riflemen were crack shots: reports state that many of them could hit a circle 7in (17.8cm) in diameter at 250yd (228m). Washington decided to increase their number in 1777 by drawing 100 men from each existing brigade. In 1778 each line-infantry battalion was ordered to create a light-infantry company, armed with smoothbore muskets, that could be brigaded in a larger light-infantry unit if necessary (Hess 2008: 22). Although the light-infantry brigades were part of the regular US Army in theory, the units were disbanded each autumn and reorganized in springtime.

During the War of 1812, four rifle regiments were raised to strengthen the US Army's regular light-infantry force; rifle companies were attached to different line-infantry regiments in action. All four regiments were disbanded in 1821, but individual rifle companies remained in the organization of each infantry regiment. According to the *Abstract of Infantry Tactics*, published in 1830, each regiment consisted of a light company, six line-infantry companies and one rifle company (US War Dept 1830: 10).

In 1843 the 2d Cavalry Regiment was redesignated as a rifle regiment for a year. When the war with Mexico broke out in April 1846, Colonel Jefferson Davis reorganized the unit, lobbying President James Knox Polk to approve his selection of the M1841 'Mississippi' rifle rather than flintlock weapons. Another rifle-armed regiment was approved by the US Congress in 1847, with half of its troops mounted and half rifle-armed foot soldiers. The Regiment of Voltigeurs and Foot Riflemen served for only one year and had an artillery detachment as well.

This M1803 rifle was manufactured at the Harper's Ferry Armory; note the wooden barrel rib, a field addition. One of the very first assignments of the new armoury was to supply 15 short rifles and replacement locks with interchangeable parts for the Lewis and Clark Expedition (1803–06), but it is not clear which type of rifle they took from the Harper's Ferry stocks. It is obvious that the new military rifle's production started only after the acceptance of the design in December 1803, but it is also proven that at least some prototypes were present at Harper's Ferry. The first possibility is that the expedition rifles were shortened and re-bored 1794 Contract rifles. The second possibility is that they were newly made prototype M1803 rifles. In a letter which arrived at Harper's Ferry on 14 March 1803, Secretary of War Henry Dearborn ordered Joseph Perkin, superintendent of the Harper's Ferry Armory, to produce the rifles for the Lewis and Clark Expedition (Carrick 2008: 1–2). Captain Meriwether Lewis arrived at Harper's Ferry two days later and supervised production of the rifles, reporting to President Thomas Jefferson on 8 July that he shot the rifles and was satisfied with the results. There are no known rifles from the expedition existing at the time of writing. Whichever possibility is true, Lewis surely had a significant impact on the design of the first US Government-made US military rifle. (Rock Island Auction)

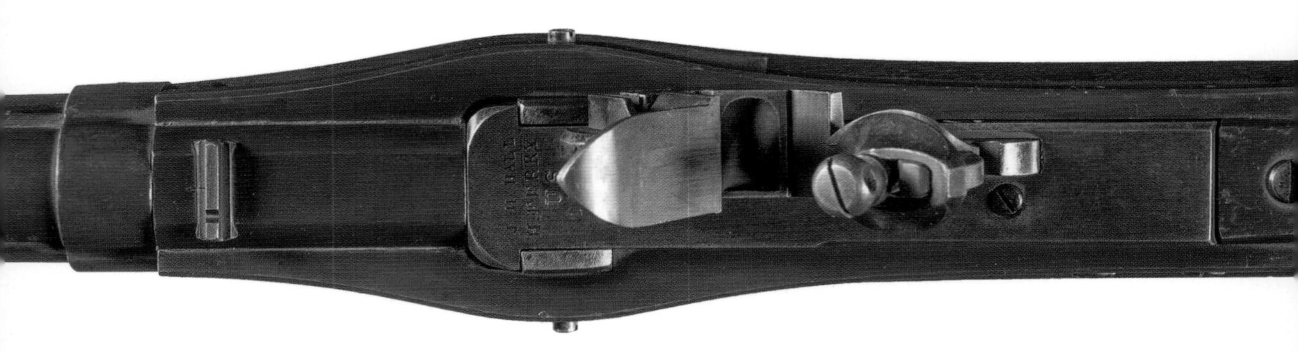

A top-down view of an 1838-dated Hall M1819 flintlock rifle. The M1819 was perfectly suited to light-infantry tactics, with a rate of fire of two or even three rounds per minute. It could be loaded in any position, even behind cover. Although the system of manufacturing, proofing and inspection was well organized, failures sometimes occurred, especially if the Hall flintlock breech-loaders were issued to less regulated troops. One such occasion involved a dragoon company serving in Florida in 1837–38: at the first discharge, five rifles were damaged, with a further eight more being broken in the second volley, sending splinters flying in all directions (Smith 1977: 217). The weapons were returned to Harper's Ferry and Hall personally inspected and reproofed them. All the barrels and receivers were undamaged. The failure was caused by the wood – steamed walnut – becoming brittle with age, and the blow-out of gases at the joint of the barrel and breech block damaging the wooden stock. Avoiding the use of steamed walnut and redesigning the stock easily resolved the problem. (Rock Island Auction)

US rifle training

During the American Revolutionary War, the drill manual of the Continental Army was written by a Prussian captain, Friedrich Wilhelm Ludolf Gerhard Augustin von Steuben. Accepted in 1779, his manual focused only on the exercises of the line infantry. The American rifleman typically knew everything about shooting his rifle, so the manuals seldom gave information about how to load the patched ball in the rifle. The preference for rifled over smoothbore small firearms was clearly understandable as on the frontier, accuracy was more important than rate of fire. Being a good shot did not equate to being a formidable soldier, however. As William Duane commented in his work *A Hand Book for Riflemen*: 'although as marksmen, the American rifle men surpass all others; in what regards discipline and the strength and confidence arising from discipline, they are inferior to the rifle men of other nation' (Duane 1812: 9). This paradox arose from the American riflemen's unfamiliarity with European-style drill. Duane's suggestion was therefore to train the rifle corps following the European method: 'Rifle corps should on their first instruction in military discipline pass through the ordinary drill and movements of a company, and afterwards of a battalion of infantry' (Duane 1812: 37).

Did this mean that the American riflemen were to employ line-infantry tactics? The answer is yes and no. Light infantry had to be trained to fight as line infantry, but also had to be able to move rapidly through rough terrain for long periods, carrying their rifles loaded so they were ready for action (Duane 1812: 38–39). In contrast to the regular infantry pace of 76 per minute, the rifleman had to be able to achieve 90 or even 120 paces per minute.

Duane wrote one of the earliest descriptions of loading the American military rifle, explaining what determined whether a rifleman would use cartridges or loose powder and ball:

> Riflemen are never required to fire with cartridges but when in acting in close order, which though it often happens, it is not precisely their province in action. Whenever it is practicable, riflemen will load with powder measure and loose ball. They must be first taught to load and fire with cartridge like other infantry; after which the principal instructions for recruits will be how to load with loose ball, and fire at the target. (Duane 1812: 43)

US RIFLE AMMUNITION

Until the 1840s, military rifles shot lead balls that were cast with hand moulds, usually gang moulds capable of casting 12–24 bullets in one operation. The balls were smoothed in rolling barrels or by shaking them in a bag for several minutes. There were very few bullet-making machines available. Using eight ten-gang musket-ball moulds and operated by two men – one pouring the molten lead and the second turning the handles – Daniel Pettibone's moulding machine was capable of making 300 balls per minute. Invented in 1818, Pettibone's machine was not adopted by the US Government (Thomas 1997: I.100–01).

According to *Military Pyrotechny for the use of The Cadets of the United States Military Academy*, published in 1832, 66.7 grains of musket powder were measured in a paper tube rolled from a trapezoid paper sheet 4.3in (10.9cm) long with 2.2in (5.6cm) and 3.3in (8.4cm) sides (Thomas 1997: I.118–19). The size of the ball was .504in (12.8mm). These instructions do not mention the use of patching. Based on the size of the bullet and powder charge, the author reasons that these instructions were written for the 1792 and 1794 contract rifles. The unpatched ball was oversized compared to the land-to-land diameter of the rifle, suggesting that these bullets had to be hammered into the muzzle.

The US Army standardized its cartridge specifications in 1835 (Thomas 1997: I.99). The cartridge used by the M1803 and M1814 rifles held a .525in (13.3mm) lead round ball and 100 grains (6.48g) of fine rifle powder. The fit of the ball was tight as it was rammed into the bore wrapped in a square piece of thin muslin, leather or bladder patch. Rifles were loaded with an undersized patched ball. The ball had to be smaller than the land-to-land diameter to be able to enter the bore easily. Wrapping the ball in a greased patch created a gas tight seal, eliminated the windage between ball and bore and created a mechanical fit so the rifle could spin the bullet to stabilize its flight. The lubrication of the patch kept the fouling of powder residue soft, so the rifle could be shot several dozen times before it got fouled.

The construction of the paper cartridge started with the patch tied over the ball, leaving a projecting end. The assembly was dipped into molten tallow. The paper tube was rolled on a mandrel. The length of the paper trapezoid sheet was 4.25in (10.8cm); the longer side was 4in (10.2cm), while the shorter side was 2.25in (5.7cm). The case of the cartridge was 'choked' over the tied projection to hold the bullet-and-patch assembly in the end of the paper tube, and it was fastened by two turns and a double stitch (OM 1841: 181). The construction of the cartridges was done entirely by hand: one worker was responsible for one stage. A 19-man group included ten to roll the cylinders, one to fill them, four to fold and four to bundle. A team was able to make 10,000 cartridges during a ten-hour shift. According to the *Ordnance Manual* of 1850, the charge of all rifle cartridges was 75 grains (4.86g) of fine rifle powder (OM 1850: 243).

The cartridges of the Hall rifle were constructed similarly. The ball was .525in (13.3mm) in diameter. The land-to-land diameter of the bore – .520in (13.2mm) – was slightly smaller than the bullet diameter, so when the gas pressure forced it into the rifling, it created a mechanical fit in the .01in-deep (0.25mm) 16-groove rifling. The diameter of the hole chamber was .545in (13.8mm), so loading the cartridge was easy and did not require the use of force (OM 1841: 99). According to the *Ordnance Regulations* of 1834, the charge was 100 grains (6.48g) of rifle powder, including approximately 10 grains for priming (Thomas 1997: II.133). The cartridges of the M1833 North-Hall percussion carbines held a lighter charge of 75 grains (4.86g) of powder as priming was not necessary. The *Ordnance Manual* of 1841 still mentioned a 100-grain-charged rifle cartridge (OM 1841: 178), but the 1850 edition reduced the charge to 75 grains (4.86g) in the case of rifles, suggesting that most of the Hall rifles were converted to percussion, and also supporting the adaptation of the M1841 rifles (OM 1850: 243). The manuals did not give instructions on how to make cartridges for breech-loading rifles. According to the *Ordnance Manual* of 1841, the height of a flintlock rifle cartridge was 2.5in (6.4cm) (OM 1841: 183), while according to the second edition of 1850, the cartridge length was reduced to 2.1in (5.3cm) due to the lighter powder charge (OM 1850: 248).

The ingredients for making the necessary calibre ammunition – powder, balls, sheets of paper for cutting the trapezoids and threads – were also carried by US Army units. Learning to roll cartridges was a necessity for every soldier.

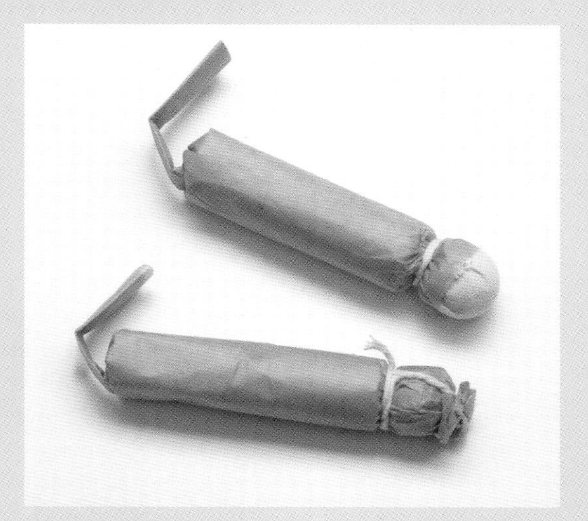

Two reproduction US rifle cartridges used for the Hall breech-loaders. The upper cartridge holds a patched ball; the paper tube is tied to the 'choked' patch. The paper tube of the lower cartridge has been closed with a double stitch above and below the ball; the end of the paper case was folded (Thomas 1997: I.109; Hoyem 2005: 56). (Author's collection)

The loading method for charging the rifle with a paper cartridge did not differ from musket drill; the only difference was the introduction of a 'wad' over the powder. Duane did not properly explain the nature of this element, and no description can be found in any of the contemporary American manuals.

Rifle target practice came after the musketry drill. Firing started at 50yd (46m), and as the shooters perfected their skills the distance was improved in stages to 300yd (274m). According to Duane, the best target was at least 5ft (1.52m) in diameter, circular, with painted circles, so the shooter could observe his errors and correct them. All target practice was to be carried out offhand, to avoid getting used to supports. The soldier had to understand how to vary the amount of powder charge depending on the distance to the target, and had to learn to ram the ball down firmly without crushing the powder grains. Duane mentions also the use of 'plaister', a piece of greased flannel, fustian or leather, to facilitate the passage of the ball into the bore, and clean the residue of the previous shot (Duane 1812: 44).

Duane also described an interesting style of target practice employed by some Pennsylvanian riflemen. The target could be a hat or similarly sized object placed on a fence post. Using the 'Indian pace', the company moved around the target in a circle, keeping the target on their left side, while constantly firing without stopping the movement – a useful tactical element for light troops (Duane 1812: 44). Rapid movement was an important element of light-infantry tactics. The ability to retreat quickly to rally points was vital as the rifles lacked bayonets, therefore the light troops were extremely vulnerable if forced into close combat.

US rifle tactics

Light-infantry soldiers could select individual targets even when firing from close-order formations: 'Light infantry and rifle companies will be particularly instructed in file-firing, as in open order, as this will be the mode usually adopted by them in the field. In this fire, the soldier will not be required, as heretofore prescribed, to aim direct to the front, but will be allowed to select his object to the right, left, or front' (US War Dept 1825: 251).

The US War Department's *A System of Tactics*, published in 1825, introduced two new firing positions. All soldiers had to learn to load and fire kneeling and lying down. Kneeling was used in close-order combat, when firing volleys, and when fighting in chain formation, for increased accuracy and to reduce the shooter's profile. Loading the rifle when lying down was not easy, however. The soldier had to roll onto his left side, bring back the rifle to have the lock near his chest for priming the pan, and then roll onto his back to load the cartridge into the bore (US War Dept 1825: 252–54).

Major General Winfield Scott's *A System of Tactics*, published in 1834, described the loading methods for rifle-armed troops in eight commands, a simplified procedure compared to that followed by the line infantry:

US RIFLE ACCESSORIES

Irregular riflemen of the American Revolutionary War and the War of 1812 seldom had centrally issued equipment, therefore the militiaman carried exactly the same accoutrements and accessories for his rifle that he used for hunting game.

As the regularization of the rifle troops progressed, the standard accessories were listed in the *Ordnance Manual* of 1841: the bullet mould, a leather pouch for the balls and cartridges and a copper powder flask with valve and adjustable spout for measuring a powder charge between 85 grains (5.5g) and 100 grains (6.48g) (OM 1841: 96). These accessories hung on a buff leather waist belt, attached by brass hooks. Each rifle was accompanied by a rifle sling, a brush and a vent-hole pick (OM 1841: 141). All other accessories were the same as those used by the line infantry. The lack of a line-infantry cartridge box and the existence of a powder flask with adjustable valve suggests that loading the military rifle with loose powder and patch separated from the ball was just as common as using paper cartridges. This loading method was especially important for irregular light troops as the calibre of their firearms could vary considerably, making it impractical to supply balls or cartridges from central sources. According to the *Ordnance Manual* of 1841 the best small-arms caps, designated No. 1 caps, were the so called 'split caps', charged with the priming compound, a white powder mixed from 100 parts of mercury-fulminate and 60 parts of saltpetre (OM 1841: 211).

1. LOAD!
One motion.
1214. Carry back the right foot, making a half-face to the right, turning on the left heel, let fall the firelock, seizing it with the left hand at the swell, the elbow resting against the left side; the right hand quits its hold, placing the thumb against the top of the hammer.
2. *Open* – PAN!
One motion.
1215. The pan is pushed open with the right thumb; the right hand seizes the cartridge with the three first fingers, carries it to the mouth, which tears off the end, whence it is brought close to the pan.
3. PRIME!
One motion.
1216. The priming is shaken into the pan; the pan is shut by the third and little finger; the right hand then slides behind the cock, and holds the small of the stock between the third and little finger, and ball of the hand.

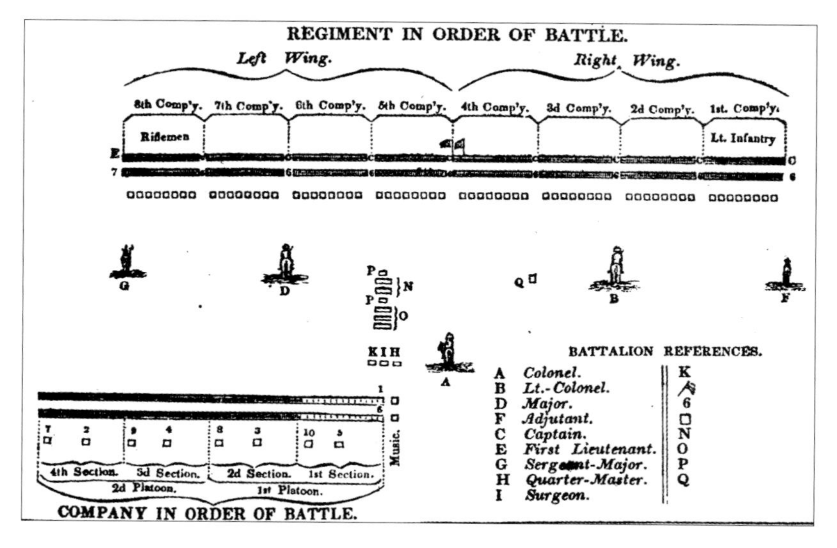

This diagram is taken from *Abstract of Infantry Tactics* (US War Dept 1830: 10). The most important close-order combat formation was the two-ranks-deep deployed battalion line. The tallest men were placed on the flanks, but the soldiers of the rear rank were supposed to be taller than those of the front rank. This facilitated loading and firing by both ranks simultaneously. Behind the two ranks stood the 'rank closers': commissioned or non-commissioned officers (US War Dept 1830: 37). While a European regiment had 3,000–4,000 men, a US regiment was the size of a European battalion, and was also called a 'battalion' during manoeuvres. According to the 1825 edition of *A System of Tactics*, it consisted of ten companies, eight 'battalion companies' (traditional line infantry) plus a light company and the rifle or grenadier company, termed 'flank companies'. If a grenadier company was present in the organization it was placed on the right flank of the deployed line (US War Dept 1825: 7–8). Each company was divided into two platoons, and each platoon was divided into two sections. The *Abstract of Infantry Tactics*, intended for militia, did not mention grenadiers, leaving only the rifles and light troops as flank companies. (Author's collection)

4. *Cast* – ABOUT!
First motion.
1217. The soldier fronts, bringing the right foot to its original position; the rifle is brought, with the barrel outwards, (sliding it with care through the left hand,) to the ground, the butt placed between the heels, the barrel between the knees, which must be sufficiently bent for that purpose; the left hand takes hold near the muzzle, the thumb stretched along the stock.
Second motion.
1218. The cartridge is put into the barrel, and the ramrod seized with the fore-finger and thumb of the right hand.
5. *Draw* – RAMROD!
One motion.
1219. The ramrod is drawn by the right hand, the left quits the rifle and grasps the rod, the breadth of a hand from the bottom, which is sunk one inch [25.4mm] into the barrel.
6. *Ram* – CARTRIDGE!
One motion.
1220. The cartridge is forced down by both hands; the left then seizes the rifle near the tail pipe; the soldier stands upright, and seizes with the thumb and fore finger, the small end of the rod.
7. *Return* – RAMROD!
One motion.
1221. The rod is drawn out, and returned by the right hand, which remains with the ball resting on the head of the ramrod – elbow square.
8. *Shoulder* – ARMS!
First motion.
1222. The left hand carries the rifle to the right shoulder, turning the guard outwards, the right receiving it in its proper position at the small.
Second motion.
1223. The left hand is carried quickly to the left thigh. (Scott 1834: 173–75)

One important difference compared to the Austrian method was the lack of a wooden mallet for starting the ball, as the patched ball was not as tight-fitting as in the case of the *Jägerstutzen* rifles. A new manual for percussion firearms was not published until 1855, demonstrating the low

'Mississippi' rifles at the battle of Buena Vista (opposite)

The artwork shows the moments just after the Mississippi riflemen led by Colonel Jefferson Davies fired their last volley at the charging Mexican cavalry at Buena Vista on 23 February 1847. The riflemen are equipped with M1841 'Mississippi' rifles and the rifleman's standard accoutrements based on the *Ordnance Manual* of 1841: cartridge pouch made of upper shoe leather attached to the flask and pouch belt made of buff leather with two loops, white duck haversack and wooden canteen. The first rifles of the type did not have bayonet lugs, nor were they equipped with socket bayonets, so the soldiers armed themselves with large Bowie knives. This plate shows the moment when many of the soldiers have already dropped their rifles and drawn their Bowies to initiate close combat against the halted Mexican light cavalry.

US
US

tactical importance of changing the ignition system of the infantry firearms. The modernization of Scott's *A System of Tactics* became inevitable when the rifle-musket was accepted for general service. William Hardee's *Rifle and Light Infantry Tactics*, published in 1855, implemented the change. Capping the nipple became the last operation of the procedure, while priming was the first in the case of flintlocks.

BRITISH RIFLE-ARMED TROOPS

The riflemen fielded by Britain before 1775 were a mixture of foreigners hired for service in a particular campaign, British regulars temporarily assigned from line-infantry regiments and – in North America – irregular provincial troops fighting in the service of the British Crown, who provided their own weapons. The rifles purchased by Colonel Prevost were used during the expedition against Fort DuQuesne, the campaign against Fort Ticonderoga and the attack on Fort Carillon, all conducted in 1758 (Bailey 2002: 13–16).

British interest in rifled bores increased dramatically at the outbreak of the American Revolutionary War. Propaganda articles published in Britain praised the effectiveness of rifles and the marksmanship of the American riflemen. Captain Ferguson volunteered to be the commander of a trial company for field testing. After seeing action in 1777–78 (see page 64) his unit was disbanded and his men were sent back to their regiments, some of them keeping their rifles and green jackets.

During the early part of the French Revolutionary Wars, Britain continued to employ foreigners armed with rifles, and it was not until the opening years of the 19th century that British regulars once again carried rifled firearms. Even then, Britain's rifle-armed specialists – notably of the 95th Regiment, eventually deploying three battalions, and the 5th and 7th battalions of the 60th Regiment – continued to be armed with a variety of German and other patterns of rifles alongside the Baker. The fame won by British riflemen earned them a permanent place in the British Army in the years after the final defeat of Napoleon.

Late-18th-century riflemen of the Electorate of Hessen-Cassel, carrying the *ältere kurhessische Jägerbüchse*. The Electorate provided a significant force of riflemen for the British Army during the American Revolutionary War. The backbone of the Hessian light-infantry force were the *Jägerbataillon* and the *Schützenbataillon*, both consisting of 623 men. Each soldier was armed with a rifle, equipped with a *Hirschfänger* attaching to a rail on the right side of the muzzle. Although central specifications for these rifles did not exist, they followed the common design of German *Jäger* rifles. The calibre of the old models was 15.5mm, with a 79.5cm barrel, rifled with seven grooves (Götz 1978: 174–75). From left to right, this image shows an officer, a non-commissioned officer, a hornist and a *Jäger*. The hornist was equipped with a bugle horn, an important feature of the light troops. As these units were often detached from the main body of the army and their soldiers fought in loose combat formations, a louder, more distinctive communication method than oral commands was necessary for maintaining continuous contact while in action. (Anne S.K. Brown Military Collection, Brown University Library)

British rifle training

The British Army was a latecomer in raising its own rifle troops, but still contributed much to using the rifle. Ezekiel Baker was one of the most important military thinkers at the time of the French Revolutionary Wars and Napoleonic Wars, publishing his experiences on rifle shooting on a regular basis and contributing important rules of thumb concerning accuracy that remain applicable today: 'After the trigger is pulled, keep the rifle firm to the shoulder, till the ball strikes the target at 100 yards [91m] (...) A rifleman should practise to pull the trigger, with a wood-driver in the cock, till he can fire off his piece without starting, or shaking the muzzle of his rifle, or blinking ...' (Baker 1806: 8). He advised not to train with the rifle on a support: 'The rifle recruit must at first be taught to fire at the target without a rest, for if he accustoms himself to make use of a support, he will rarely fire true without one' (RER 1803: 13).

Equally important was accurately judging the target's distance up to the rifle's furthest tactical range: 'A rifleman, to judge of his distance, should be in the habit of stepping his ground from 1 to 300 paces [0.75–225m], or any other distance that may be thought proper (...) By this continual practice, he will learn to measure the distance with his eye to a tolerable certainty at any time' (Baker 1806: 8).

According to the official regulations, the target practice was to be held at four distances: 'The firing will be divided into four ranges for practice; the 1st for the recruits at 90 yards [82m], the second at 140 yards [127m], the third at 200 yards [182m], and the 4th at 300 yards [273m], beyond which no established practice is to go' (RRC 1800: 114). The target had to provide enough space to be hit by 'flyers' – hits on the target that were out of the group and therefore errors:

> The round target or circle of wood 4 feet [1.22m] diameter, painted white, with three circles in black: – the first circle is 4 inches [10.2cm] from the centre of the bull's eye, which is not to be more than 1½ inch [3.8cm] diameter, the 2nd at 9 inches [2.29cm] from the centre, and the 3rd or out circle at 15 inches [3.81cm]; each ring is to be 2 inches [5.1cm] broad. (RRC 1800: 114)

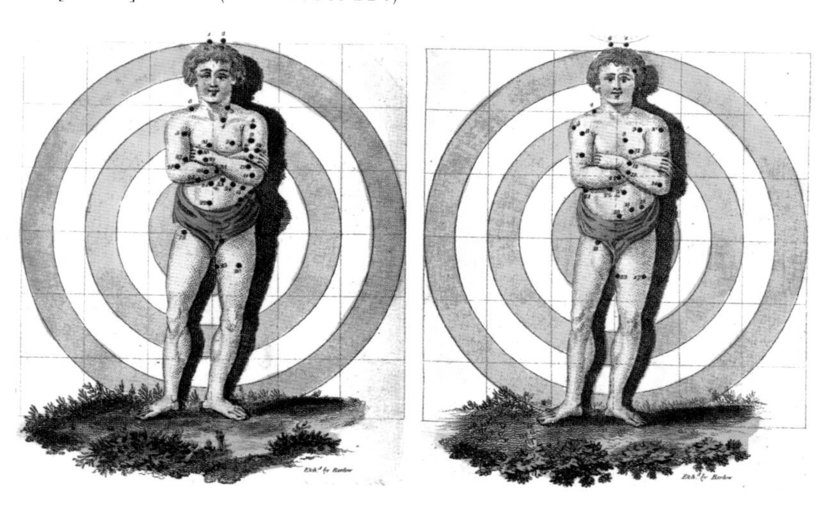

100yd (91m) group (left) and 200yd (183m) group (right) of the Baker rifle, published in Baker 1806. Practising at the shooting range in ideal circumstances was important for learning the basics, but to become a master of rifle fire demanded more: 'I would recommend a young rifleman when he can fire well at 200 yards [182m], in calm weather, to practise in windy, and all sorts of rough weather. (...) I have found 200 yards [182m] the greatest range I could fire at any certainty. At 300 yards [273m] I have fired very well, at times, when the wind has been calm. At 4 and 500 yards [364m and 455m] I have frequently fired; and I have some times struck the object' (Baker 1806: 9). (Author's collection)

The recruits were split into three classes:

> The 1st class, or bad shots – the 2nd tolerably good – and the 3rd marksmen. (...) The 1st class, or the bad shots, are always to be fired at the two 1st ranges, the 2nd class occasionally at them, but in general at the 3rd range, but never at the last; and the marksmen occasionally at the 3rd, but in general at the 300 yards [273m], or last range of distance. Each man on all days of practice, will fire six rounds, and those six rounds are to be at one given distance for that day. (...) Captains will keep in their possession a firing-book in which the two monthly rolls are marked, and also the daily work of the target. (RRC 1800: 116)

There were serious debates about whether these ranges were good or not:

> It has been frequently objected to the rifle corps in the British service, by military men, that their ranges of practice were much too short; for, say they, a rifle should begin where a musket ceases to be of use; (...) the most powerful inducement to the adaptation of long ranges, will be shew how much more a rifleman will be benefited than by sticking to the shorter ones, we only wish to recall in mind, that from a musket at 300 yards [273m], not one shot in 100, or we may say 300, would if fired at a single man as the object, take effect; when on the contrary, with a rifle we may take, at least, one in five; but more likely, in skilful hands, one in three as a fair average. (Beaufoy 1808: 193)

Baker offers detailed instructions of loading for maximum accuracy. The ball had to be placed in the centre of the patch on the muzzle, otherwise the rifle threw the ball to the side. The ball would not be hammered to meet the powder charge but instead was pushed down 'with as little bruising as possible' (Baker 1806: 11) to avoid crushing the powder and deforming the ball. The ball had to be rammed firmly against the powder because if an air gap remained between them, the barrel could burst (Baker 1806: 11; RER 1803: 13). The pan had to be closed before the ball was inserted, as the airflow blew the fine rifle powder through the touch hole. The hammer had to be in the half-cock position to prevent it from falling on the frizzen and setting off the charge during loading. The touch hole had to be obstructed by a picker to prevent powder from spilling in the pan. The picker was used to force some powder into the touch hole to secure ignition (Baker 1806: 19–21).

British rifle tactics

The experiences of the first years of the American Revolutionary War were controversial. It was quickly understood that being a good marksman was only a small part of the skills a soldier needed to fight a disciplined regular force such as the British Army. Such an occasion at Throg's Neck, New York (12 October 1776), was described by William Dansey, captain of the Light Company of the 33rd Regiment of Foot (Bailey 2002: 22).

BRITISH RIFLE AMMUNITION

During the Seven Years' War, British regular troops were accompanied on campaign in North America by irregular provincial troops, many of whom carried rifles. In a June 1758 letter to Brigadier-General John Forbes, Colonel Henry Bouquet wrote that many of the troops from Pennsylvania, Virginia and Maryland carried rifles and bullet moulds; they required FF-grade powder and lead bars rather than bullets (Bailey 2002: 16). Mentioning the supply of lead bars indicates that the calibre of the rifles varied, while the reference to fine FF-grade powder suggests that these troops used small calibres, probably smaller than the German *Jäger* calibres.

At the beginning of the American Revolutionary War, Colonel William Faucitt, having been tasked with recruiting Electoral battalions in Germany for service with the British Army, provided information about German powders, and requested stronger propellant for his P1776 rifles. The fine powder was manufactured by Messr Bridges, Eade & Wilton in Surrey. Each P1776 rifle was furnished with enough powder for 200 shots. One 100lb (45.4kg) barrel of the 'superfine double-strength' (SDS) powder cost £7.10.0. The complete shipment to the British Army serving in America consisted of 1,000 rifles, 90 barrels of rifle powder, 18½ tons (18.8 tonnes) of carbine balls, 10,000 flints and 200 formers for cartridges. No information is available on the weight of the charge, but the cartridge formers and the use of finer powder suggest a 3 dram or 82-grain (5.316g) charge, including 5–10 grains (0.324–0.648g) for priming. There is no indication of using a patch (Bailey 2002: 26–27 & 34). Captain Patrick Ferguson requested accoutrements and accessories for the rifles bearing his name: bayonets and scabbards, rifle slings, riflemen's belts, powder horns and twelve-gang bullet moulds.

Ezekiel Baker suggested a charge of 118 grains (7.6g) of rifle powder for his rifle (Baker 1806: 20), but the Board of Ordnance reduced it to 96 grains (6.2g). The slow twist-rate required a higher charge for accuracy, so after the Napoleonic Wars it was increased to 110 grains (7.1g). For target work the soldier loaded the rifle with loose powder and a patched ball, but for rapid firing he loaded the rifle with a paper cartridge holding a naked 355-grain round ball. During the Napoleonic Wars, two cartridges were produced for these rifles. The 'forced ball' cartridge held a .596in (15.1mm) patched round ball, while the 'running ball' cartridge designed for rapid fire held a .615in (15.6mm) naked ball (Bailey 2002: 151). This method – having two different ball sizes for the same rifle – made logistics more complex. Information about the patch is seldom found. Baker's suggestions were the following: 'A ball forced down too hard, or yet too easy; I never found to go so true, as when properly fitted. The ball with its patch should fit air tight, or it will not have the desired effect' (Baker 1806: 10). Further information about the patch is offered by other sources: 'By the patch is understood a small piece of greased leather, &c. which is put round the ball before driving it down, to fill up the interstices of the grooves, which would, without this precaution, accession too great windage. The requisites in a patch are, strength, elasticity, and that it be of equal thickness in every part' (Beaufoy 1808: 165). Blank and ball cartridges were differentiated by the colour of the paper: '... Blank Cartridges for Exercise are on every occasion to be made up, exclusively, in Blue Paper' and 'Ball Cartridges are to be made up in Brown or Whited-brown Paper' (GOR 1811: 90). The yearly allowance of cartridges for a rifleman was 'sixty rounds of ball cartridges, three Flints per Man, of which Proportion 2-thirds are issued in the Spring and the remainder in the Autumn' (GOR 1811: 70). This is not much compared to today's standard, but this volume of ammunition was in line with other major powers' light-infantry cartridge allowances.

The Brunswick rifle's 2¼-dram (3.99g) powder charge was wrapped in a blank-paper cartridge; the bullets were issued to the soldiers wrapped in greased calico patches with a black band marking the position of the belt to facilitate loading. Soldiers were also issued cartridges holding 2½ drams (4.43g, 68.4 grains) of musket powder and standard, unpatched musket balls for rapid firing. George Lovell also made an important improvement to ball production, proposing that they should be made on David Napier's compressing machine, rather than cast in the traditional way. The pressed balls proved superior in the trials, so Napier's invention was purchased for military use in 1839 (Hare 2019).

Reproduction of a British carbine cartridge. The paper is tied close above and below the ball. The powder column is also closed with a string with the tail of the cartridge twisted (Hoyem 2005: 6). (Author's collection)

Dansey noted that although his opponents were marksmen who enjoyed superior cover, after exchanging fire for several hours at a range of 200yd, none of his men had been hit.

Rifles at Kings Mountain, 7 October 1780 (previous pages)

The plate shows the engagement between William Campbell's Patriot backwoodsmen and Captain Patrick Ferguson's Loyalist militia at Kings Mountain, South Carolina. The Patriots were mainly armed with long rifles, and were well accustomed to hunting and fighting in the woods. They were better able to utilize the advantages of the terrain than the Loyalist militia, which mainly deployed into close-combat formations. The rifle accessories carried by the frontiersman in the foreground include a large powder horn with nipple pick and pan brush hanging on a leather strap, and a possibles bag. Long rifles were not equipped with bayonets, therefore most of the frontiersmen armed themselves with large hunting knives or tomahawks.

As the Patriot backwoodsmen close in on their opponents, one Ferguson rifle-armed soldier has survived and can be seen loading his weapon. Although the use of the Ferguson rifle at Kings Mountain is not mentioned in the sources, indirect sources suggest that it was present in the engagement: the sources indicate the presence of riflemen in Ferguson's contingent: several British carbine bullets were recovered from the battlefield, and a flintlock hammer supposed to be the hammer of a Ferguson rifle was found at the location (Roberts & Brown 2011). Proving the Ferguson rifle's presence will be the task of future researchers.

There are not many sources on the overall quality and accuracy of the British muzzle-loading rifles used in the American Revolutionary War, but the journalist William Clachar stated that they were fine firearms (Bailey 2002: 26). Clachar noted that one gentleman was able to place six out of eight shots in an area no bigger than the crown of his hat at a range of 150yd (137m); at 400yd (366m) he shot within 18in (45.7cm) of the mark.

During the American Revolutionary War Ferguson's command would be deployed in pitched battles, but also in the 'small war' activities common to all rifle-armed troops. His unit set sail from Portsmouth on 26 March 1777 and disembarked on 26 May in New York, and half a month later was heading towards New Jersey as part of a light-troop detachment. Ferguson fought a typical 'small war', leading his unit through many small engagements until he was badly wounded at the joint of the right elbow on 11 September 1777 at the battle of Brandywine. Valuable information about the tactics of Ferguson's riflemen is found in a January 1778 letter to his brother, in which Ferguson explained that because his riflemen were able to load and fire their weapons while lying down, he did not lose a single man during the battle despite being engaged on six occasions (Bailey 2002: 49).

Parts of Ferguson's former detachment took part in a night raid on Brigadier General Anthony Wayne's sleeping army at Paoli on 20–21 September 1777. The Ferguson rifles were withdrawn from the

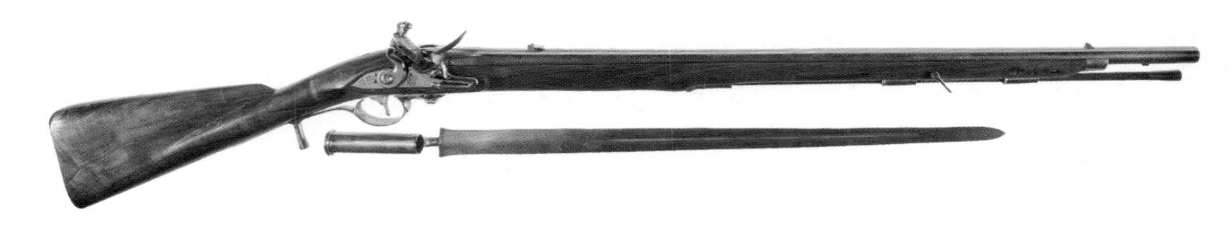

A reproduction Ferguson rifle. The Ferguson firing procedure started with putting the hammer into half-cock, opening the pan, and then opening the breech with a complete turn of the trigger guard. Then the soldier bit the end of the cartridge, primed the pan and closed the frizzen. The bullet and powder were then introduced into the chamber and the breech was closed with one full turn. After cocking the hammer, the rifle was ready to fire. The key advantage of the Ferguson rifle was not its high rate of fire, but the silence of loading behind cover or in a lying position (Bailey 2002: 20). (Rock Island Auction)

BRITISH RIFLE ACCESSORIES

The most important tools for the Baker rifle were stored in the patch box: the ball screw, the wiping eye and the torque bar, to be inserted into the hole of the ramrod for a better grip. The soldier carried a powder horn or powder flask on a green cord, a brass powder charger (1810–14), a ball bag to carry 30 balls, a wooden mallet to start the bullet (issued only with the first batch of rifles), a cartridge pouch with a wooden box for 12 cartridges and a tin compartment for 24 cartridges, a lock cover made of oiled leather or cloth and a bayonet and scabbard. There are indications that Baker-armed soldiers carried a bullet mould likely to cast the special-size 'running ball' projectiles described on page 61 (Bailey 2002: 145–51).

various light-infantry units on 21 February 1778. Ferguson stayed in America, and after a long recovery received his next assignment in the autumn of 1780 to lead a party of Loyalist militia soldiers into North Carolina. He met his fate at Kings Mountain on 7 October 1780 when the camp of his retreating unit of 800 musket- and rifle-armed militia was surrounded by Campbell's 900 rifle-armed backwoodsmen (see page 64). Ferguson failed to fortify the camp, and failed to set up outposts. The backwoodsmen surrounded the mountain. The Loyalist forces tried to repel them with bayonet charges, but the Kentucky rifles delivered an accurate and deadly fire on the defending forces. The backwoodsmen conquered the slopes and broke into the lines of Loyalists, killing and wounding 388 and capturing 716 of them. Ferguson was killed in action, hit by eight bullets.

In the effort to defeat Napoleon and secure Britain's overseas possessions, the British Army's newly formed rifle units fought in campaigns all over the world in the first years of the 19th century, from Montevideo to Copenhagen, but would win their lasting fame in the Iberian Peninsula. This was where the Baker rifle really came into its own.

British riflemen could load their Baker rifles with paper cartridges like the line infantry: 'When a company or battalion of riflemen is to act with closed ranks and files, the same regulations which are given to infantry in general serve for them. And before the soldier is instructed in the manoeuvres of light troops, he must be taught how to hold himself, to march, face, wheel, as in regular infantry' (RER 1803: 1). On these occasions, pre-rolled 'naked' ball cartridges were used. The loading procedure was the same as in the case of muskets: the soldier took a

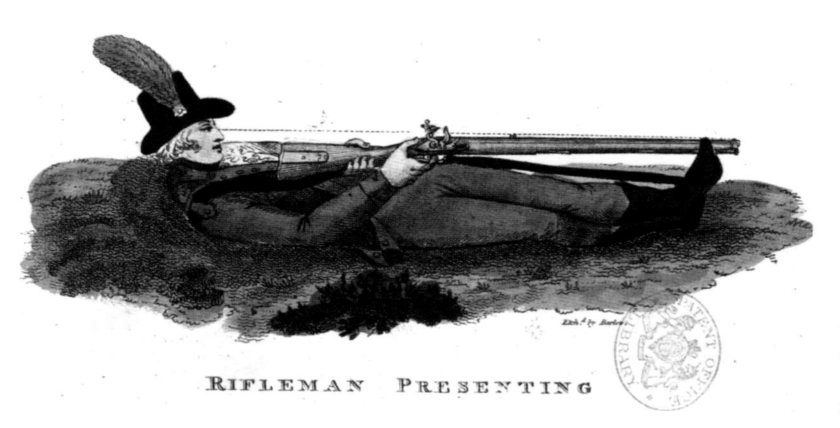

The standing, kneeling and prone positions were well used by light troops all across Europe, but the British Army was the only one to use the back position for further support of the arm as suggested by Baker. He knew that a light trigger pull was vital for accuracy: 'The trigger should not draw so hard as to alter the direction of the rifle in firing. (…) I do not mean air-triggers, but the triggers with common pull, as used by the 95th regiment; nor gun over heavy, nor very light' (Baker 1806: 15). To achieve this pull he installed a detent or fly in the tumbler to help the sear jump over the half-cock notch: 'and the locks which have a fly or scape, in the tumbler, to prevent the seer catching at the half cock ...' (Baker 1806: 15). Not all locks were supplied with this little part installed, but most of them were later modified by Nock and Baker in 1801. This illustration is taken from Baker 1806. (Author's collection)

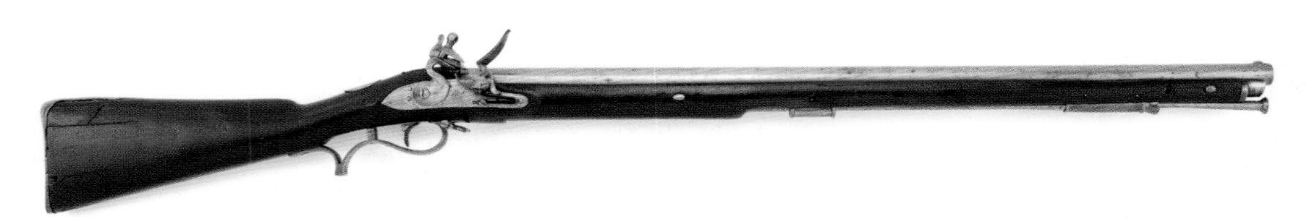

This Baker rifle is in the collections of Springfield Armory; it looks like the P1800, but the hammer has a swan neck. Note that the bayonet rail is missing from the barrel. The Baker rifle's barrel was fixed with three keys, which aided easy removal of the bore for cleaning. The cleanness of the bore was an important factor in achieving accuracy, meaning that it was advisable to wipe the bore frequently during practice sessions to remove the powder residues. The Baker rifle needed to be cleaned immediately after use, as the black-powder residues absorbed moisture from the air and created an acidic compound that destroyed the rifling: 'we do believe, more harm is done to a gun, by suffering it to remain in a dirty state after use for a week, than with fair usage would accrue from a month's shooting' (Beaufoy 1808: 202). If the barrel got rusty, fine emery powder was to be added to the greased side of the patch. This method removed the rust in a few shots. (USNPS Photo, Springfield Armory National Historic Site, SPAR 8142)

cartridge from his cartridge box, bit the end of the paper case, primed the pan, lowered the rifle's butt to the ground between his heels, grasped the rifle at the muzzle, loaded the powder and the ball into the bore and rammed it down with the ramrod. This was the procedure when a high rate of fire and harmonized loading rhythm was required; otherwise, riflemen loaded their rifles with patched ammunition.

FRENCH RIFLE-ARMED TROOPS

Having eschewed the military rifle for many decades, the French Army adopted the weapon in response to the unusual irregular tactics encountered in North Africa following the French occupation of Algeria in 1830. The French found that pacification of this vast territory was difficult due to the activities of large bodies of irregular indigenous opponents. The French Army was not trained or equipped to counter their opponents' fighting practices; the indigenous fighters avoided open clashes, used hit-and-run light tactics extensively and manoeuvred beyond the musketry range of the smoothbore-armed French troops. These experiences in Algeria strengthened the need for regular, fast-moving, rifle-armed light troops – a type of infantry the French Army lacked. In 1838, Ferdinand Philippe, Duke of Orléans, received permission to raise a rifle company that was expanded into a battalion in 1840. The prince visited Prussia to gain experience of rifle troops. In 1840, based on the success of the new unit, nine more battalions were established that created the core of the French Army's rifle troops.

French rifle tactics

New open-order combat tactics were developed for this new type of French infantry. The Algerian fighters often occupied ditches and trenches for cover and let the slow-moving French soldiers approach them across open terrain. Manoeuvring in close-order formations proved suicidal for the French, who were forced to develop new skirmishing tactics. According to the contemporary military theorist Friedrich Engels, the French skirmishers worked in groups of four, each group in a line with five paces (3.75m) between men and up to 40 paces (30m) between groups. The NCOs positioned themselves ten paces (7.5m) behind their sections, with the officers, accompanied by a bugler and four soldiers, 20 (15m) or so paces to the rear. The availability of cover was given priority over the exact shape of the formation, and was used to move stealthily towards the enemy before suddenly appearing at close quarters (Engels 1959: 88–90).

Increasing the speed with which troops could move and attacking in open order offered a significant advantage when fighting against a rifle-

FRENCH RIFLE AMMUNITION AND ACCESSORIES

Capitaine Henri-Gustave Delvigne's chamber-breech rifle originally fired a 29.3g lead round ball united with 96 grains (6.22g) of musket powder in a paper cartridge. The M1846 *carabine à tige* rifle fired a 46.6g cylindro-conical bullet propelled by a 4.5g (69.4-grain) powder charge, the reduction being necessary to compensate for the rifle's heavy recoil (Mordecai 1861: 162). The muzzle velocity was 310m/sec (Mordecai 1861: 232). The sleeve of the paper cartridge held the projectile upside down, with the bullet's point towards the powder charge. The cartridge was lubricated on the outer side. The soldier opened the other end of the cartridge with his teeth, poured the charge into the bore, turned the cartridge and inserted the projectile with the bullet still wrapped in the greased-paper patching. This method prevented lead build-up in the grooves and also improved the seal of the gases of the burning powder by filling the windage between bore and bullet.

The most important accessory of the French rifles was the 'Yataghan' sword-bayonet with brass handle. The 59cm single-edged blade was steel, fullered on both sides. The bayonet was replaced by the Chassepot bayonet in 1866. The tip of the Thouvenin rifle's ramrod was countersunk to protect the form of the bullet while hammering the projectile on the pillar. The standard accessories of the Thouvenin rifle included a ball puller for removing a dead charge and a patch puller for removing cleaning patches from the bore. Both tools fitted to the threaded end of the ramrod.

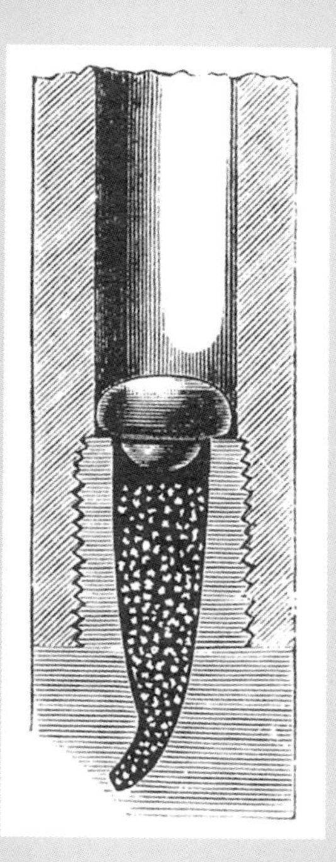

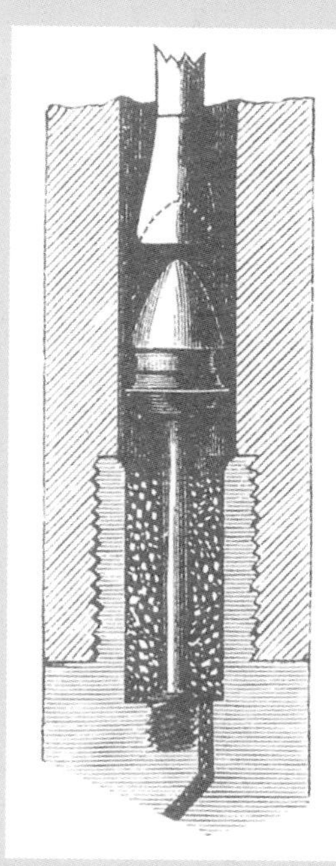

Two views of Delvigne's chamber breech (left and middle) and one of Thouvenin's pillar breech (right), taken from Belházy 1892. The first image shows how the undersized bullet was loaded into the bore; the second image shows the bullet after it has been hammered and upset on the chamber mouth by the heavy thrusts of the ramrod. The third image shows how the undersized conical bullet is hammered on the pillar with heavy thrusts of the ramrod, expanding its base into the rifling. Note that the tip of the ramrod was countersunk; this was necessary to protect the point of the bullet. (Author's collection)

armed enemy: the time period moving in the fatal zone of rifle fire could be shortened significantly, while employing a dispersed combat formation reduced the target area offered by the attacking units. The French Army developed a completely new system of attack – shock tactics. Infantry units were deployed into extended or open order well outside the effective range of rifle fire and then moved quickly using the newly adopted *pas gymnastique* – 165–180 steps (139–151m) in a minute – to approach the enemy until reaching a distance where all shots counted. The attacking unit now fired a limited number of volleys and charged the enemy formation with fixed bayonets. Ironically, the general adoption of the rifle as a general infantry arm promoted the use of the bayonet (Németh 2017a: 115).

The evolving appearance of Prussia's rifle troops, 1760–1846, is depicted in this illustration. (Museum Wolmirstedt, Inv. No. KG_4754.09)

IMPACT

The early rifles' performance

It is not easy to find contemporary sources describing the capabilities of the early military rifles with scientific precision. Analysing the few records that do exist and modern shooting tests aids our understanding of the advantages and drawbacks of these rifles in combat.

RIFLE vs SMOOTHBORE MUSKET

The advantages in accuracy conferred by the rifle must be measured against the capabilities of the smoothbore musket, and there are three

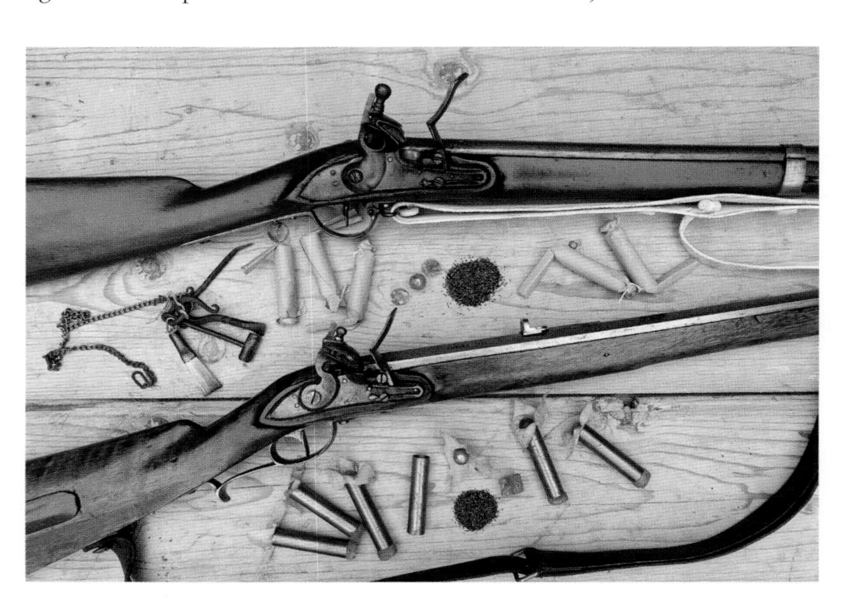

The 17.6mm-calibre M1798 *Infanteriegewehr* (above), a smoothbore weapon issued to the Habsburg infantry, was 149.3cm overall with a barrel length of 111.7cm and weighed 3.74kg, while the 13.9mm-calibre M1807 *Jägerstutzen* rifle (below) was 104.7cm overall with a barrel length of 66.5cm and weighed 3.35kg. (Author's collection)

At left, 100 paces (75m) group size of the M1798 musket fired offhand: all shots were in the target, with many in the body. At right, 150 paces (112m) group size of the M1798, with only one hitting the target, missing the body. (Author's collection)

important measures to follow in experimental archaeology shooting projects. First, the firearm must be an original in perfect condition or a perfect replica. Second, the projectile has to be an exact reproduction of the service bullet. Third, the powder charge must be measured to reach the documented muzzle velocity.

The author's range tests were carried out using an original M1807 *Jägerstutzen* rifle and a reproduction M1798 *Infanteriegewehr*, a smoothbore weapon, matching these three criteria. The cartridges of the M1798 musket were loaded with 11g (169.8 grains) of Swiss 1.5Fg black powder – approximately equivalent in terms of corn size to 19th-century military musket powders – and a 15.9mm ball. The average muzzle velocity of the charge was 497m/sec, matching contemporary sources such as those written by Hermann Cotty and William Duane. The charge of the M1807 rifle was a 14.1mm patched ball and 4.375g (67.5 grains) of Swiss 2Fg powder – a close equivalent in terms of composition and corn size to the military rifle powder used by the Habsburg Army – resulting in an average muzzle velocity of 400m/sec. In the case of the M1807 we do not

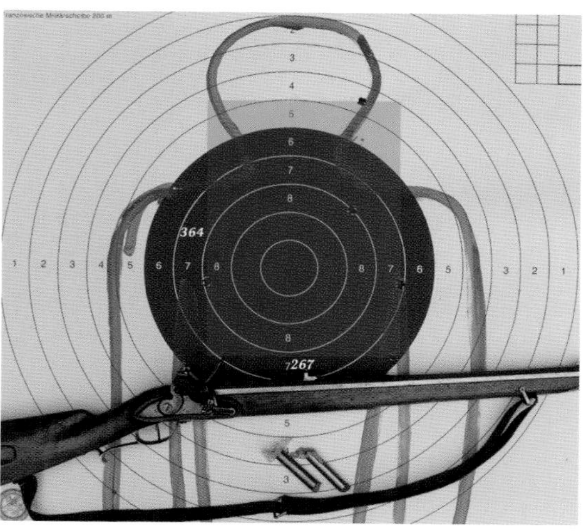

100 paces (75m) group size (left) and 150 paces (112m) group size (right) of the M1807 rifle fired offhand. All shots were in the target and all in the body. (Author's collection)

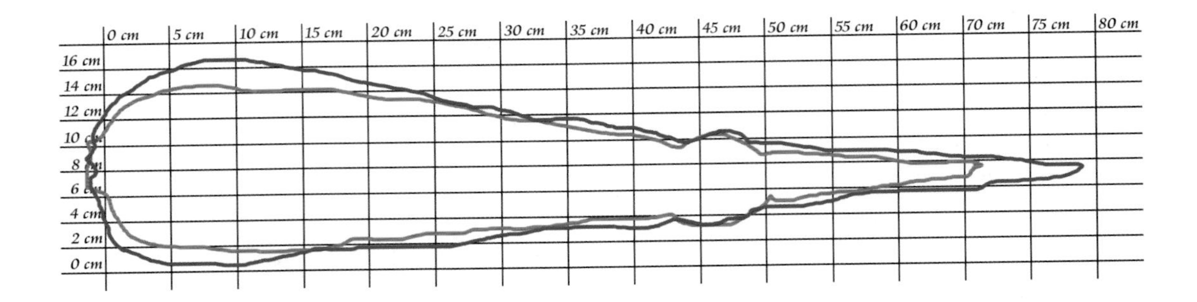

During the tests, killing power was verified by shooting the projectiles into ballistic gelatine at a range of 50m. Regardless of the charge, the bullets made the same temporary and permanent cavities, with the rifle ball penetrating 71cm and the musket ball 79cm into the gelatine block. Here, the temporary cavities of the two firearms, recorded at 980 frames per second, have been redrawn (blue for the smoothbore musket and red for the rifle; the depth is along the x-axis and the width is along the y-axis). Both bullets share similar terminal ballistics. (Author's collection)

have contemporary information about the muzzle velocity, so the closest-matching commercially available powder was selected based on samples taken from an original cartridge.

The accuracy of both firearms was tested by shooting offhand, wearing full contemporary uniform and using the properly constructed cartridges and loading methods, at ranges of 100 paces (75m) and 150 paces (112m). At 75m, the M1798's best five-shot group out of six shots could be covered with a 27.7×64.9cm rectangle, while the M1807's five-shot group could be covered by a 12.5×15.9cm rectangle with all shots in the target. At a range of 112m, however, the situation changed significantly. The M1798 had only one hit on the 80×80cm target out of five shots, while the M1807 placed all five in a 36.4×26.7cm rectangle – well within the area of a human chest.

The M1798 needed an average of 35 seconds to load and fire each shot, while the M1807 needed 60 seconds, including more time for accurate aiming. Loading the smoothbore musket was straightforward during the whole shooting session, with the substantially undersized ball being easy to ram down even a fouled barrel. The most time-consuming parts of the process were removing the cartridge from the cartridge box and drawing the ramrod from the stock. By contrast, the rifle needed significantly more effort to load. Finding the components of the charge in the cartridge box was time consuming, while starting the bullet in the muzzle with the mallet also took valuable time. Shooting in each relay with each arm commenced with fresh, sharp flints but even if the flint was brand-new, the first misfires arrived between the eighth and the tenth shot.

INTERPRETING THE HISTORICAL RECORD: PRUSSIAN RIFLES

The most reliable source we have for the early Prussian flintlock rifles (Scharnhorst 1813) mentions two rifle types used in range tests on 6 October 1810. One, named the *Alte Corpsbüchse*, was a 15.52mm-calibre weapon 104.6cm long and weighing 4.25kg. The other – described as 'a rifle made in the gun repair facility of Berlin' – was in 14.7mm calibre; it was 117.7cm long and weighed 4.33kg (Scharnhorst 1813: 88). The first rifle was loaded with a 15.4mm-diameter ball weighing 21.74g, while the second used a ball weighing 19.23g and which was, at 14.8mm in diameter, slightly larger than the bore.

Scharnhorst indicates two loading methods. Undersized balls were in most cases loaded with lubricated linen or leather patching, while the balls closely matching the rifles' calibre were loaded without patching. The *Alte Corpsbüchse* variant described by Scharnhorst was tested by ten *Jäger* firing 100 shots at each of four different ranges. There was no powder charge set for the shooters, who were free to select the one to which they were accustomed; their charges varied between 2.92g and 7.3g. The target area at all distances was a pinewood screen 1in (26.16mm) thick, 6ft (1.88m) high and 24ft (7.53m) wide, with a section 4ft (1.26m) wide and 6ft (1.88m) high in the middle nominated as the true target area (Scharnhorst 1813: 93). The rifles' accuracy and penetration were tested both with loose powder and patched ball and also with cartridges.

Accuracy of the *Alte Corpsbüchse* with *Pflasterkugel* and *Patronenkugel* cartridges

Type of charge	Number of shots	Distance	Hits	Penetrations
Patched ball of normal calibre loaded with loose powder (*Pflasterkugel*)	100	150 paces (113m)	68	68
	100	200 paces (150m)	49	47
	100	300 paces (225m)	31	not visible
	100	400 paces (300m)	20	not visible
Paper cartridge with smaller patched ball (*Patronenkugel*)	100	150 paces (113m)	51	not visible
	100	200 paces (150m)	26	not visible

A quick evaluation of these results suggests two things: first, the *Jäger* involved in these trials were poor shots (or their rifles were inaccurate, or both); and second, their charges were extremely weak, not being able to penetrate the wooden screen above a distance of 300 paces (225m). The paper cartridges were loaded with a ball 0.65–0.78mm smaller than the standard patched ball to facilitate loading. The ball was wrapped in a square patch with the four edges 'choked' with a string and tied into the paper case of the cartridge. The wrapped ball was than dipped into molten tallow for lubrication. Even though the balls were substantially undersized and the patches were lubricated, the rifles were still slow to load. The time for firing the ten shots varied between 18 and 26 minutes. The tests also demonstrated that after approximately 20 shots the rifles were so fouled that they had to be cleaned before shooting could continue (Scharnhorst 1813: 91).

REPUTATION AND REALITY: THE US AND BRITISH RIFLES

The organic development of the American long rifle resulted in probably the best flintlock rifle of the late 18th century, and its performance was recognized by the contemporary European gunmakers as well. Colonel George Hanger of the British Army serving in North America was one of many commentators who praised the American rifleman and his weapon

Dating from 1815–17, this engraved and silver-accented long rifle has been signed by the gunmaker Jacob Kunz of Philadelphia, Pennsylvania. (Rock Island Auction)

(Kauffman 2005: 24). The long rifle's lighter projectile had a flatter trajectory than that of the European *Jäger* rifle, making long-distance aiming easier with less elevation. Contrary to popular misconception, however, the longer barrel did not itself contribute to greater accuracy. The longer distance between the rear and front sights aided accurate sight arrangement, while the lighter bullet and powder charge resulted in less recoil. These were important features, but within the contemporary tactical and hunting distances – up to 200yd (183m) – both rifles types' performance was adequate.

During the Napoleonic Wars, the best British riflemen were also capable of exceptional accuracy. William Surtees, a former rifleman who served in the British Army during the Peninsular War and for many years afterwards, mentioned the exceptional shooting skills exhibited by one of his officers: 'This officer (Major Wade) was one of the best shots himself that I have ever seen. I have known him, and a soldier of the name Smeaton, hold the target for each other at the distance of 150 yards [137m], while the other fired at it, so steady and accurate was their shooting' (Surtees 1833: 42).

THE RIFLE'S TACTICAL IMPACT

The muzzle-loading rifle could not itself revolutionize warfare as it shared the key problem of all such firearms: the complexity of the loading procedure limited the rate of fire and always offered the enemy an opportunity for a bayonet charge. The rifle's increased effective range was useful in itself, but up until the end of the muzzle-loading era musketry it alone – including rifle fire – was not enough to decide the outcome of the fight. Even if the rifle's effective range was tripled compared to that of the smoothbore musket, close-order combat formations and volleys were still essential for the rifle-armed soldier to be able to inflict serious damage on the enemy, and to resist enemy charges. This system could only be changed

The .424-calibre percussion rifle by R.M. Wilder of Coldwater, Michigan (see page 32), was also put to the test over the same distances as the Austrian military rifle. This heavy-barrel target or picket rifle was made in the 1850s; the fast twist-rate of the rifling requires a conical bullet. Although this rifle was made primarily for civilian target shooting, firearms of this type were widely used during the American Civil War for sharpshooting and sniper duties. The first loads the author used were various patched round ball combinations, but none of them grouped well regardless of the thickness and material of the patch or the quantity and quality of the powder charge. The most accurate load was 50 grains (3.24g) of 2Fg Swiss black powder and an 18.25g elongated, skirted (Minié-style) bullet. The 75m offhand group (left) could be covered with a 70×64mm rectangle, while at 112m (right) an 89×64mm rectangle was sufficient to cover the best five shots of a six-shot relay. Although it is rather unfair to compare a civilian target rifle with a military rifle, this test clearly proves the superiority of elongated bullets over patched round balls at longer ranges. (Author's collection)

with the introduction of fast-firing breech-loading firearms. Consequently, the rifle's range advantage over the smoothbore musket did not have a decisive impact on infantry tactics. It was the increased firing rate of the breech-loader that would transform rifle tactics.

Rifle-armed troops and rifles played a crucial part in pioneering and developing the tactical solutions that were subsequently utilized by troops armed with breech-loading rifles during the second half of the 19th century. The general acceptance of the rifle in the mid-19th century effectively abolished the barriers between light and heavy infantrymen by introducing light-infantry principles to all types of foot soldier. Building on the tactical innovations implemented by the French in North Africa (see page 66) and harnessing the potential of the M1841 *Zündnadelgewehr*, the Prussian Army developed new rifle tactics to be implemented by the entire infantry arm using open-order combat formations. Operating the M1841's bolt action needed only five movements, many fewer than any muzzle-loader (Németh 2017b: 163–66); the rifle could fire 6–7 shots per minute in the hands of the average soldier.

The new Prussian system of *Feuertaktik* (rifle combat) was based on the skirmish line. The *Schwarmlinie* (chain) had support troops and reserves just as in the times of the Napoleonic Wars, but the purpose of this formation was now substantially changed. It was not only responsible for initiating the combat, but also for fighting the battle by providing an

The breech of the Ferguson rifle in open position. Modern-day shooting tests were carried out by Ernest E. Cowen and Richard Keller in 1999. They discovered that the plug screw was subject to fouling and clogging if the rifle was loaded with loose powder, when the excess powder was used for priming the lock. With the use of properly lubricated bullets and breech plug, and regular cleaning of the upper section of the threads in the breech, the Ferguson rifle was capable of firing any number of shots, but without them it clogged easily, demonstrating the vulnerability of the system (Bailey 2002: 220–21). (© Royal Armouries XII.11209)

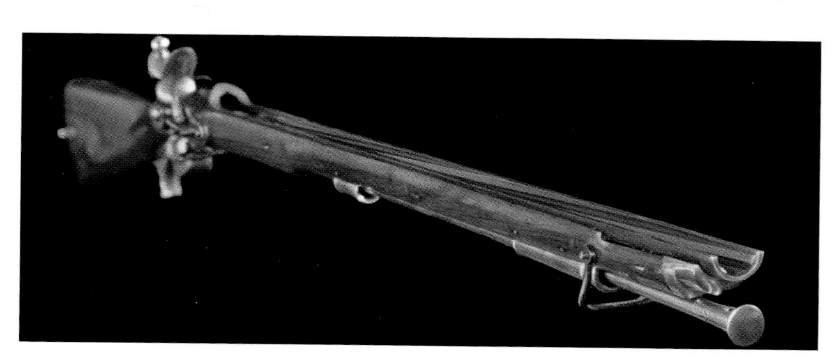

A P1801 Baker rifle with its barrel cut to show the rifling. The rifling had seven grooves with a twist-rate of one turn in 120in (305cm). Note the slow twist-rate and deep grooves. The Baker rifle's brass front sight was soldered to the bore. The rounded top of the sight offered an excellent sight picture even in low light conditions. The rear block sight had a hinged leaf for larger distances. (© Royal Armouries XII.2443)

Dated 1815 and bearing Harper's Ferry Armory markings, this M1803 flintlock rifle is from the second production run. The barrel was still 33in (83.8cm) long, but the ramrod was equipped with a brass tip to protect the rifling. The new rifles' wood screws were already being machine-made not hand-forged, demonstrating the development of production methods. The length of the barrel was increased to 36in (91.4cm) from June 1815. (Rock Island Auction)

increasing intensity of rifle fire. The basic element of these tactics was to select the position of the line with skill. To utilize the benefits of the fast-firing M1841, it was advisable to hand the tactical advantage of attack to the enemy. The Prussian infantryman could use cover offered by the terrain and could fire his rifle in any position. As the enemy approached the Prussian formation, the support troops – and, if it was necessary, the reserves – were pushed forward into the line to increase the density of fire. When the enemy approached to a distance less than 150 paces (113m), the *Schnellfeuer* (rapid fire) began. Each Prussian soldier loaded and fired his breech-loading rifle as fast as he could. The M1841 was equipped with a socket bayonet, but in Moltke's system of tactics the bayonet charge was the very last element of the combat. The fight had to be decided with the rifles (Németh 2017b: 161–62).

This coloured lithograph by Franz Gerasch shows an Austrian *Jäger*, 1848. Principles such as the use of camouflage colours for uniforms originated with the rifle-armed troops of the patched round ball period. The grey uniforms worn by the Habsburg Army's *Jäger* and the green of the Prussian Army and British Army light troops were the first modern military uniforms. The comfortable loose legwear and the practical accoutrements to facilitate the rifle-loading procedure were new paradigms. The *Jäger* in the picture is armed with the M1842 *Kammerbüchse*, but wears the accoutrements and accessories related to the M1842 *Jägerstutzen* rifle: the sword-bayonet, the powder horn hanging on a green cord, and the ramrod visible on his right side. This is most probably a mistake on the part of the artist who mixed up the two types of soldiers. The rifleman is fighting in open-order combat formation; he is probably a member of the chain, using rocks as cover and support for his rifle. (Author's collection)

CONCLUSION

It is often stated that the main reason why the rifle did not become the standard-issue infantry firearm earlier was its slow rate of fire. The present author does not agree with this statement, as the rifle could be loaded as quickly as the smoothbore musket if paper cartridges holding unpatched balls were used, and still delivered better accuracy than the smoothbore weapon. The main factor stopping the rifle from replacing the smoothbore musket was military logistics: the rifle was too expensive to replace if damaged or lost, and was too fragile for close combat.

The Swiss M1851 *Eidigenössischer Stutzer* or *Feldstutzer* rifle, a percussion muzzle-loader, was the first small-calibre military rifle accepted for service. It had a practical long-range rear sight, making it particularly suitable for mountain warfare. (Author's collection)

The final improvement of the muzzle-loading military rifle was a parallel invention by Joseph Lorenz of the Vienna Arsenal and the Englishman Henry Wilkinson. In 1853, they both designed a bullet with deep compression grooves without a hollow skirt. The principle was the same as in the case of the expanding projectiles: the bullet was undersized when loaded, but was upset into the rifling by the force of the burning gases. With this invention the muzzle-loading rifle reached its limits. Its effective range against a single soldier was increased to 200–250m, and improvements in manufacturing methods allowed the machine production of sufficient rifles to equip all soldiers with such firearms. Now that 2–3 shots per minute was achievable, the rate of fire and the ease of loading and maintenance of the rifle finally reached the level of the old smoothbore weapons.

In the 1850s all of the key military powers adopted percussion rifles for the entire army. The British accepted the P1851 Minié rifle, and later reduced the calibre to .577in with the P1853 Enfield rifle firing the .568in paper-patched Pritchett ball. The Habsburg Army adopted the 13.9mm M1854 rifle firing a 28g Lorenz-style compression bullet. The US Army accepted the M1855 Springfield rifle in .58 calibre firing a modified version of the British bullet, the 510-grain (33g), .575in (14.6mm) Burton-Minié ball. The French Army was the only major army that retained the old large-calibre bore and adopted the M1857 Minié rifle in 18mm calibre

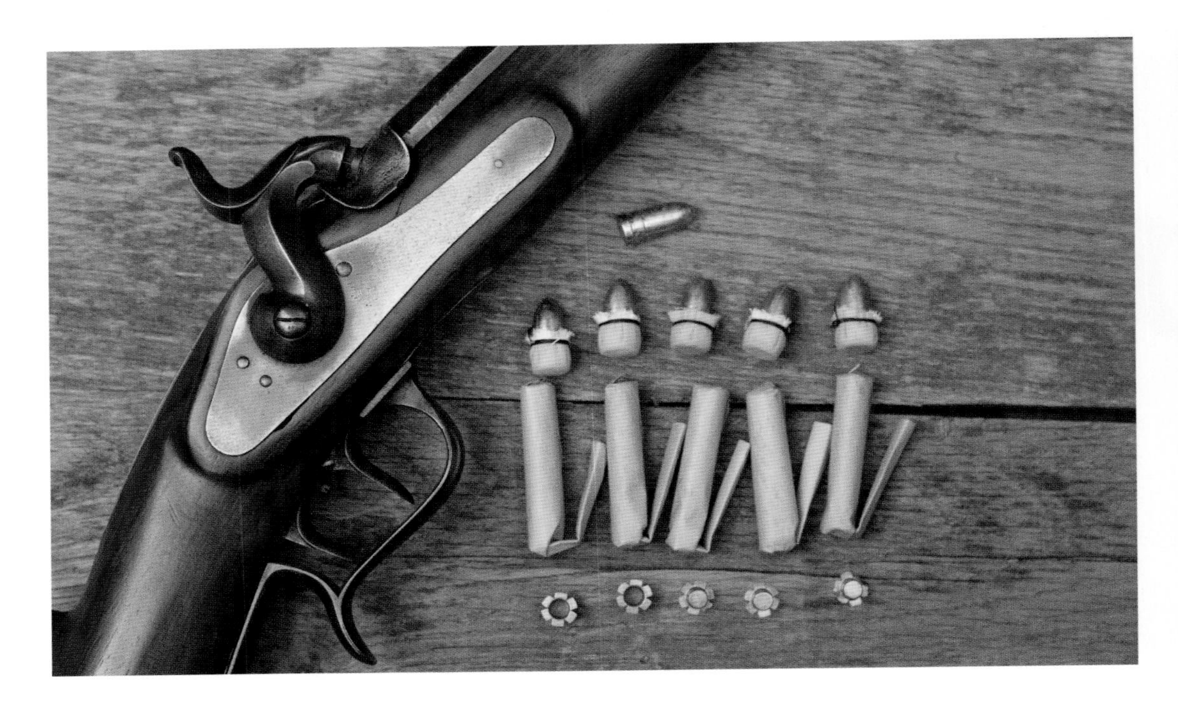

The Swiss M1851 rifle's first cartridge contained a patched 10.4mm conical bullet weighing 16.5–17g and 60 grains (3.89g) of fine rifle powder. The muzzle velocity of the bullet – 440m/sec – was the highest of all contemporary military rifles, resulting in an extremely flat trajectory. (Author's collection)

firing a 17.6mm Minié projectile weighing 48g. All the infantry percussion rifles accepted for service in the 1850s had one important drawback compared to the smoothbore musket, however: the muzzle velocity of the rifle projectiles had to be limited to 280–380m/sec, as at higher pressure the soft lead bullets tended to jump the rifling and lose accuracy. The smoothbore musket shot its loose-fitting round balls at a muzzle velocity of 450–500m/sec. On the other hand, the weight of the conical ball was greater than that of the musket ball. Firing a heavier bullet at a lower velocity resulted in a highly curved trajectory that travelled high above the head of the soldier at ranges up to 200m. To address this, it would be necessary to develop a lighter, more advanced, aerodynamic bullet. In the 1850 and 1860s efforts were made by Joseph Whitworth and others to harden the alloy by adding tin to the lead, and redesign the rifling profile.

Also, using the new long-range sights was not an easy task for the average soldier. Loading and discharging a smoothbore musket in the line required much less understanding, thinking and training than was necessary to learn the proper use of the rifle and light-infantry tactics. With the general issue of percussion firearms, the world's armies faced a new challenge: the recruits of all armies, even the humblest foot soldier, had to understand the basics of ballistics and had to attain the same level of knowledge as the most skilled riflemen of the Napoleonic era. This entailed a different approach to training, and a different approach to education before soldiers joined the army. Musketry – and ballistics – became scientific areas of study, and in order to educate NCOs and officers, musketry schools such as those at The Hythe, Vincennes and Bruck an der Leitha were established all over Europe. The minimal requirements for each recruit increased dramatically and strengthened the case for general elementary education.

BIBLIOGRAPHY

Published sources

Anonymous (1851). *Abrichtungs-Reglement für die k. k. Jäger* [Training Regulation for the Imperial–Royal *Jäger*]. Vienna: k.k. Hof- und Staatsdruckerei.

Baker, Ezekiel (1806). *Twenty-Six Years Practice and Observations with Rifle Guns*. London: Wright.

Beaufoy, Henry (1808). *Scloppetaria: Or Considerations on the Nature and Use of Rifled Barrel Guns*. London: C. Roworth.

Belházy, Jenő (1892). *A vadászati ismeretek kézikönyve* [Essential Knowledge of Hunting]. Budapest: Grill Károly.

Beroaldo-Bianchini, Natalis-Felix (1829). *Sammlung der Pläne sammt Erklärung zum Behufe der Abhandlung über die Feuer- und Seitengewehre* [Collection of Plans with Descriptions for the Purpose of the Treatise on Firearms and Sidearms]. Vienna: Hof- und Staats-Aerarial-Druckerei.

Bailey, De Witt. (2002) *British Military Flintlock Rifles 1740–1840: The Story of Melvin Maynard Johnson, Jr., and his Guns*. Lincoln, RI: Andrew Mowbray Publishers.

Carrick, Michael F. (2008). 'U.S. Model 1803 Prototype Rifle', in *American Society of Arms Collectors Bulletin* 97: 1–4. Available at https://americansocietyofarmscollectors.org/wp-content/uploads/2019/06/2008-B97-U-S-Model-1803-Prototype-Rifle.pdf (accessed 24 December 2019).

Clausewitz, Carl von, ed. & trans. C. Daase & J.W. Davis (2015). *Clausewitz on Small War*. Oxford: Oxford University Press.

Cotty, Hermann (1806). *Mémoire sur le fabrication des armes portatives de guerre* [Treatise on the Fabrication of Handheld Military Arms]. Paris: Chez Magimel.

Dirrheimer, Günther (1975). *Die K.K. Armee im Biedermeier* [The Imperial and Royal Army in the Age of Biedermeier]. Vienna: Tusch.

Dolleczek, Anton (1887). *Geschichte der österreichischen Artillerie von den frühesten Zeiten bis zur Gegenwart* [History of the Austrian Artillery from the Earliest Times until Today]. Vienna: Kreisel & Göger.

Dolleczek, Anton (1970). *Monographie der k.u k. öster.-ung. blanken und Handfeuer-Waffen* [Monograph of the Imperial and Royal Cold Weapons and Firearms]. Graz: Akademische Druck- u. Verlagsanstalt. First published in 1896.

Duane, William (1812). *A Hand Book for Riflemen*. Philadelphia, PA: Self-published.

Engels, Friedrich (1959). *Engels as a Military Critic*. Manchester: Manchester University Press. A collection of articles originally published in the 1860s.

Flanagan, Edward R. (2008). '1792 and 1807 Contract Rifles', in *American Society of Arms Collectors Bulletin* 97: 30–38. Available at https://americansocietyofarmscollectors.org/wp-content/uploads/2019/06/2008-B97-1792-and-1807-Contract-Rifles.pdf (accessed 24 December 2019).

Gabriel, Erich (1990). *Die Hand- und Faustfeuerwaffen der habsburgischen Heere* [The Handheld Firearms of the Hapsburg Army]. Vienna: ÖBV.

Götz, Hans-Dieter (1978). *Militärgewehre und Pistolen der deutschen Staaten 1800–1870* [Military Rifles and Pistols of the German States]. Stuttgart: Motorbuch Verlag.

Gumtau, Carl F. (1834–38). *Die Jäger und Schützen des Preußischen Heeres* [The Jägers and Riflemen of the Prussian Army]. Three volumes. Berlin: Mittler.

Hapsburg, Archduke Charles von, trans. D.I. Radakovich (2009). *Principles of War*. Ann Arbor, MI: Nimble Books LLC. Kindle Edition.

Hardee, William J. (1855). *Rifle and Light Infantry Tactics; for the Exercise and Manoeuvres of Troops when Acting as Light Infantry or Riflemen*. Philadelphia, PA: Lippincott, Grambo & Co.

Hare, Arthur (2019). 'The Brunswick rifle'. Available at http://www.researchpress.co.uk/index.php/firearms/british-military-longarms/brunswick (accessed 13 April 2020).

Hess, Earl J. (2008). *The Rifle Musket in Civil War Combat*. Lawrence, KS: University Press of Kansas.

Hoyem, George A. (2005). *The History and Development of Small Arms Ammunition, Volume 1*. 2nd Edition. Oceanside, CA: Armory Publications.

Kauffman, Henry J. (2005). *The Pennsylvania-Kentucky Rifle*. Morgantown, WV: Masthof Press.

Lugosi, József (1977). *Szélpuskák* [Air Rifles]. Budapest: Zrínyi Katonai.

Manningham, Coote (2015). 'Coote Manningham's Lectures to the 95th', in Townsend, Ben, ed., *Regulations for the Rifle Corps*. Bristol: Lulu.com: pp. 188–212. Originally published in 1803.

Mavrodin, V.V. & Mavrodin, Val. V. (1984). *Iz istorii otečestvennogo opyžija, Russkaja vintovka* (From the history of Russian arms, Russian rifles). Leningrad: Leningrad University.

Mordecai, Alfred (1861). *Military Commission to Europe, in 1855 and 1856*. Washington, DC: George W. Bowman.

Nagy, L. István (2013). *A császári-királyi hadsereg 1765–1815* [The Imperial–Royal Army 1765–1815]. Pápa: Gróf Esterházy Károly Múzeum.

Németh, Balázs (2017a). 'Az 1859. évi itáliai háború harcászata és a császári-királyi gyalogság tűzfegyverei' [Habsburg tactics and firearms of the 1859 war in Italy], in *Hadtudományi Szemle* 2017/3: 115–59.

Németh, Balázs (2017b). 'A porosz gyútűs puska és töltényei' [The Prussian needle-fire rifle and its cartridges], in *Hadtudományi Szemle* 2017/3: 160–73.

Németh, Balázs, Csikány, Tamás & Eötvös, Péter (2012). *A szabadságharc kézi lőfegyverei 1848–49* [Firearms of the Freedom Fight of 1848–49]. Budapest: Kossuth.

Paret, Peter (1966). *Yorck and the Era of Prussian Reform*. Princeton, NJ: Princeton University Press.

Paumgartten, Sigimundus (1802). *Abhandlung über dem Dienst der Feldjäger zu Fuss* [Treatise about the Service of the *Jäger* Infantry]. Vienna: Publisher not known.

Reilly, Robert M. (1972). 'Harpers Ferry and John Hall', in *American Society of Arms Collectors Bulletin* 26: 16–29. Available at: https://americansocietyofarmscollectors.org/wp-content/uploads/2019/06/1972-B26-Harpers-Ferry-And-John-Hall.pdf (accessed 24 December 2019).

Roberts, Ricky & Brown, Bryan (2011). *Every Insult & Indignity: The Life Genius and Legacy of Major Patrick Ferguson*. Scotts Valley, CA: CreateSpace. Kindle Edition.

Russell, Carl P. (2005). *Guns on the Early Frontiers: From Colonial Times to the Years of the Western Fur Trade*. Mineola, NY: Dover Publications. Kindle Edition.

Scharnhorst, Gerhard (1813). *Über die Wirkung des Feuergewehrs* [About the Effect of Firearms]. Berlin: Nauck.

Scott, Major General Winfield (1834). *A System of Tactics; or, Rules for the Exercises and Manœuvres of the Cavalry and Light Infantry and Riflemen of the United States*. Washington, DC: F.P. Blair.

Sitting, Heinrich (1908). *Geschichte des K.u.K. Feldjäger Bataillons Nr. 1: 1808–1908*. [History of the Imperial and Royal *Feldjäger* Battalion No. 1: 1808–1908]. Reichenberg: Offizierskorps.

Smith, Merritt Roe (1977). *Harpers Ferry Armory and the New Technology: The Challenge of Change*. Ithaca, NY: Cornell University Press.

Steuben, Friedrich Wilhelm Ludolf Gerhard Augustin von (1794). *Regulations for the Order and Discipline of the Troops of the United States*. Boston, MA: I. Thomas & E.T. Andrews.

Surtees, William (1833). *Twenty Five Years in the Rifle Brigade*. Edinburgh: William Blackwood.

Thomas, Dean S. (1997). *Round Ball to Rimfire: A History of Civil War Small Arms Ammunition*. Four volumes. Gettysburg, PA: Thomas Publications.
Unterberger, Leopold von (1807). *Wesentliche Kentnisse der Infanterie- und Cavallerie-Feuergewehre zum Gebrauch der Officiere der k.k. Oestreichischen Armee* [Essential Knowledge of Infantry and Cavalry Firearms for Use of Officers of the Imperial–Royal Austrian Army]. Vienna: Wappler & Beck.
Urban, Sylvanus (1776). *The Gentleman's Magazine and Historical Chronicle*. Volume XLVI. Available at https://ia802807.us.archive.org/34/items/thelondonmagazineorgentlemansmonth/The_Gentleman_s_Magazine_and_Historical.pdf (accessed 30 April 2020).
US Army (1841). *Ordnance Manual*. Washington, DC: Gideon & Co.
US Army (1850). *Ordnance Manual*. Washington, DC: Gideon & Co.
US Army (1834). *Regulations for Government of the Ordnance Department*. Washington, DC: F.P. Blair.
US Army Ordnance Department (1823). *Regulations for the Inspection of Small Arms, 1823*. Washington, DC: Davis & Force.
US War Department (1825). *A System of Tactics; or, Rules for the Exercises and Manœuvres of the Cavalry and Light Infantry and Riflemen of the United States*. Three volumes. Washington, DC: F.P. Blair.
US War Department (1830). *Abstract of Infantry Tactics: Including Exercises and Manœuvres of Light-Infantry and Riflemen: for Use of Militia of the United States*. Boston, MA: Hilliard, Grey, Little & Wilkins.
War Office (1803). *Regulations for the Exercise of Riflemen and Light Infantry and Instructions for their Conduct in the Field*. London: T. Egerton.

Other sources

Auszug aus dem Exercier Reglement für die k. k. Infanterie vom Jahr 1807. Vienna, 1807.
Exerzir-Reglement für die Infanterie der königlich preußischen Armee. Berlin, 1812.
General Orders and Regulations of the Army. London, 1811.
Instruction für die Frei-Regimenter oder leichten Infanterie-Regimenter, Berlin, 1783.
Instruction für sämtliche Infanterie-Regimenter und Fusilier-Bataillone. Exercieren der Schützen betreffend. Berlin, 1789.
'Instruktion über die Kenntniß und Behandlung der Preußischen Jäger-Corps Büchse', in Gumtau, Carl F. (1838). *Die Jäger und Schützen des Preußischen Heeres: was sie waren, was sie sind und was sie sein werden*, Supplement 7. Berlin: Mittler.
Reglement für die Königlich Preußische Infanterie. Berlin, 1788.
Regulations for the Rifle Corps formed at Blatchington Barracks, under the Command of Colonel Manningham. Bristol, 1800.
Report from the Secretary of War, transmitting a report of a board of officers appointed to examine the improvements in certain fire-arms; in compliance with a resolution of the Senate of 27th January, 1837. 3 October 1837, 25th Congress, 1st Session.
Report No. 1375, 63d Congress, 2d Session, Heirs or assigns of Joshua Shaw.
US Militia Acts of 2 May 1792, 8 May 1792 and 23 April 1808.
US Patent 1516X, Hall, John H., Thornton, William: Improvement in fire-arms, 21 May 1811.
US Patent 3131X, Blanchard, Thomas: Turning irregular forms, 6 September 1819.
US Patent 3686, Edward Savage, Simeon North: Improvement in fire-arms, 30 July 1844.

INDEX

Figures in **bold** refer to illustrations.